OCS Study
MMS 2005-032

Understanding the Processes that Maintain the Oxygen Levels in the Deep Gulf of Mexico

I0476164

Synthesis Report

Authors

Ann E. Jochens
Leslie C. Bender
Steven F. DiMarco
John W. Morse
Mahlon C. Kennicutt II
Matthew K. Howard
Worth D. Nowlin, Jr.

Prepared under MMS Contract
1435-01-02-CT-85080
by
Texas A&M University
Department of Oceanography
College Station, Texas 77843-3146

U.S. Department of the Interior
Minerals Management Service
Gulf of Mexico OCS Region

New Orleans
June 2005

DISCLAIMER

This report was prepared under contract between the Minerals Management Service (MMS) and the Texas A&M Research Foundation. This report has been technically reviewed by the MMS, and it has been approved for publication. Approval does not signify that the contents necessarily reflect the views and policies of the MMS, nor does mention of trade names or commercial products constitute endorsement or recommendation for use. It is, however, exempt from review and compliance with the MMS editorial standards.

REPORT AVAILABILITY

Extra copies of the report may be obtained from the Public Information Office (Mail Stop 5034) at the following address:

U.S. Department of the Interior
Minerals Management Service
Gulf of Mexico OCS Region
Public Information Office (MS 5034)
1201 Elmwood Park Boulevard
New Orleans, Louisiana 70123-2394

Telephone: (504) 736-2519 or
1-800-200-GULF

CITATION

Suggested citation:

Jochens, A. E., L. C. Bender, S. F. DiMarco, J. W. Morse, M. C. Kennicutt II, M. K. Howard, and W. D. Nowlin, Jr. 2005. Understanding the Processes that Maintain the Oxygen Levels in the Deep Gulf of Mexico: Synthesis Report. U.S. Dept. of the Interior, Minerals Management Service, Gulf of Mexico OCS Region, New Orleans, LA. OCS Study MMS 2005-032.

ABOUT THE COVER

The cover art depicts a vertical section of dissolved oxygen in the eastern Gulf of Mexico from the Mississippi River Delta to Havana, Cuba, from data collected on the research vessel *Hidalgo* cruise 58-H-4 during 25-28 June 1958. It was adapted from a drawing in McLellan (1960), which was one of the early investigations of physical oceanography in the Gulf. The basic vertical structure shown is typical for most transects across the Gulf. Dissolved oxygen values are relatively high in the upper 100-200 m, showing the influence of the air-sea exchanges of O_2 and photosynthetic inputs of oxygen. They then decrease to a minimum at depths ranging from ~350 m to 500 m in different parts of the section due to the circulation. This oxygen minimum is found throughout the Gulf and is caused both by the local oxidation of organic material within the Gulf and the transport of the Tropical Atlantic Central Water into the Gulf from the Caribbean Sea. Below the minimum, the O_2 concentrations increase with depth. No water masses are formed in the Gulf to ventilate these deep waters. Instead, the source of the ventilation of the deep Gulf waters is the transport of oxygen-rich water masses from the Caribbean Sea into the Gulf interior. The distribution also shows the depressed (raised) isolines of O_2 in the upper 800-1000 m that are associated with anticyclonic (cyclonic) circulation features.

ACKNOWLEDGMENTS

This report would not have been possible without the contributions of a large number of people. Each principal investigator (PI) contributed ideas to or authorship of portions of the text. The principal investigators, their affiliations, and their tasks are:

Ann E. Jochens	TAMU	Program Manager, PI for Tasks 1 and 3
Leslie C. Bender	GERG/TAMU	PI for Task 2
Steven F. DiMarco	TAMU	Co-PI for Task 1
Matthew K. Howard	TAMU	Data Manager
M. C. Kennicutt II	GERG/TAMU	Co-PI for Tasks 2 and 3
John W. Morse	TAMU	Co-PI for Task 2
Worth D. Nowlin, Jr.	TAMU	Senior Scientist

The assistance of Leila Belabbassi in preparation of this report is appreciated. We thank Dr. Piers Chapman, Louisiana State University, for his thoughtful comments on the document.

We appreciate the insightful comments of the Science Review Board members: Dr. Larry Atkinson of Old Dominion University, Dr. Robert Key of Princeton University, and Dr. John Morrison of the North Carolina State University. Their comments helped to improve this report.

The enthusiastic and timely support of Dr. Mary Boatman, the MMS Contracting Officer's Technical Representative, and Dr. Alexis Lugo-Fernández has been invaluable and is greatly appreciated.

Ann E. Jochens
Program Manager

TABLE OF CONTENTS

TABLE OF CONTENTS
(continued)

LIST OF FIGURES

LIST OF TABLES

ACRONYMS

AAIW	Antarctic Intermediate Water
ADCP	Acoustic Doppler current profiler
AOU	apparent oxygen utilization
BOD	benthic oxygen demand
COH	NEGOM Chemical Oceanography and Hydrography Study
Co-PI	Co-Principal Investigator
CTD	conductivity-temperature-depth sensor
DBDB2	Digital Bathymetric Data Base 2-min
DGOMB	Deep Gulf of Mexico Benthic Study (Northern Bulf of Mexico Continental Slope Habitats and Benthic Ecology Study)
DOC	dissolved organic carbon
FS	Florida Straits
ID	identifier for cruises
LATEX A	Louisiana-Texas Shelf Physical Oceanography Program; Study A
LC	Loop Current
LCE	Loop Current Eddy
MMS	Minerals Management Service, U.S. Department of the Interior
NADW	North Atlantic Deep Water (UNADW = Upper NADW)
NEGOM	Northeastern Gulf of Mexico Physical Oceanography Program
NESDIS	National Environmental Satellite, Data, and Information Service
NMFS	National Marine Fisheries Service
NOAA	National Oceanic and Atmospheric Administration
NODC	National Oceanographic Data Center
NRC	National Research Council
NPTS	number of data points
OCL	Ocean Climate Laboratory
OCS	Outer Continental Shelf
PI	Principal Investigator
POC	particulate organic carbon
QA/QC	quality assurance/quality control
R/V	research vessel
SAIC	Science Applications International Corporation
SEAMAP	Southeast Area Monitoring and Assessment Program
SRG	Science Review Group
SOOP	Ship of Opportunity Program
STD	salinity-temperature-depth sensor
SUW	Subtropical Underwater
TACW	Tropical Atlantic Central Water
TAMU	Texas A&M University
TIGER	Texas Institutions Gulf Ecosystem Research
2.3σ	2.3 sigma = 2.3 standard deviations from the mean
WOD	World Ocean Database
XBT	expendable bathythermograph probe
YC	Yucatan Channel

1 EXECUTIVE SUMMARY

1.1   Introduction

The Minerals Management Service (MMS) of the U. S. Department of the Interior awarded the contract for the "Deepwater Program: Understanding the Processes that Maintain the Oxygen Levels in the Deep Gulf of Mexico" to the Texas A&M Research Foundation in July 2002. Scientists at Texas A&M University (TAMU) conducted the research. The study area was the deepwater Gulf of Mexico, defined as that part of the Gulf with water depths of 400 m or more. Vertically the study area extended from sea surface to sea floor. Study results will assist MMS in its management of the nation's Outer Continental Shelf Leasing Program, including oil and gas leasing in federal waters of the northern Gulf of Mexico.

The objective of this study was to understand (1) the types and rates of processes occurring in the deep Gulf of Mexico that affect the levels of oxygen in the deepwater and (2) the balance that maintains this level. The objective was addressed in five tasks. First, data search, synthesis, and reanalysis of dissolved oxygen data collected from water depths greater than 400 meters throughout the entire Gulf of Mexico basin were conducted. Second, a simple box model that described the sources and sinks of oxygen in the deep Gulf was developed. Third, the effect on dissolved oxygen concentrations of the inputs from oil and gas activities were evaluated using parameters and constants determined from the data assembled. Fourth, data gaps were identified, including an evaluation of the quality of the data and how representative it is of the Gulf's basin environment. Finally, methods for filling the data gaps, including field work and modeling studies were identified.

1.2   Data Sets

Dissolved oxygen data were acquired from MMS, other state and federal agencies, national laboratories, universities, Mexican institutions, and the private sector. Two types of data were acquired: titrated measurements of water-bottle samples and measurements from oxygen sensor probes. Dissolved oxygen data from bottle samples were obtained at nearly 4000 stations from approximately 270 cruises in the Gulf of Mexico. Data from oxygen probes were obtained for approximately 1800 stations. It is not clear whether the sensor data were properly calibrated, so these data were not used in the analyses.

The dissolved oxygen data were collected from 1922 through 2001, when this study ceased data assembly. Many of the early cruises collected bottle samples throughout the full water column, while more recent cruises sampled mainly in the upper 1000 to 1500 m. Waters at or below 1500 m are not well sampled, with less than 650 stations sampling from only a few bottles at those depths. Only the decades of the 1960s and 1970s had station coverage throughout the Gulf basin.

Data quality is variable, even within cruises. This likely is due to different analysis techniques used in the older data sets and differences in the analysts' visual identification of the end points of titration, as well as poor sampling and/or laboratory analysis. All data were retained in the archive, but suspect data points were flagged as such.

1.3   Sources and Sinks

The Gulf of Mexico is a semi-enclosed sea with two channels opening into other parts of the world's ocean. The major inflow is at the Yucatan Channel, which has a sill depth that is deep enough (~2000 m) to allow the transport of oxygen rich deep source waters from the Atlantic Ocean into the Gulf from the Caribbean Sea. The Yucatan Current, which becomes the Loop Current in the Gulf, brings these waters in. The major outflow at the Florida Straits is shallow enough (~800 m) that the deeper, oxygen rich waters can mix into the interior of the Gulf, rather than flowing directly out with the Loop Current.

The sources of dissolved oxygen in the upper waters (~100-200 m) in the Gulf of Mexico are the atmosphere and photosynthesis. Wind and wave action control the air-sea gas exchange. Photosynthesis is a local process that occurs in the upper layers and depends on light levels and nutrient supplies. The source of dissolved oxygen to the deep waters is the transport and mixing of oxygen-rich water masses into the Gulf of Mexico from the Caribbean Sea through the Yucatan Channel. There is no water mass formation in the Gulf of Mexico to replenish the deep oxygen concentrations. So, the deep circulation of the Gulf and its associated mixing are the mechanisms that replenish the deep oxygen.

The major sink of oxygen in the Gulf, as elsewhere in the world's oceans, is oxidation of organic matter. The organic matter consists of living organisms, detritus from living organisms (fecal pellets, secretions, dead organisms, etc.), natural hydrocarbon seeps that are prevalent in the Gulf, continental detritus washed into the ocean through river runoff, and anthropogenic inputs (water-borne or air-borne organic pollutants as well as inputs from hydrocarbon extraction, transportation, and consumption).

The rate of oxygen consumption in the water column by organic material has not been measured in the Gulf of Mexico, and is not easily measured in the world's ocean. However, these rates are known to decrease exponentially with increasing depth. Rates have been estimated for the Atlantic, and these can be used to estimate rates in Gulf waters. However, because the coldest deep Atlantic waters cannot enter the Gulf across the sill at the Yucatan Channel, rates of consumption used for Gulf waters that are deeper than about 2000 m cannot be based on depth, but rather should be taken at comparable temperature, which in the Gulf is approximately 4°C below ~1500 m.

The oxygen minimum zone, which generally is found in the Gulf between 300-700 m, is derived from two mechanisms. First, the Tropical Atlantic Central Water with the oxygen minimum at its core brings in waters that are depleted in dissolved oxygen from processes occurring outside the Gulf. Second, the decay of organic matter that occurs in the Gulf itself further decreases the oxygen concentrations by the same processes. However, the productivity of the Gulf is not high enough to create extreme oxygen minimum zones that are seen in other parts of the world's ocean, such as in the Arabian Sea.

1.4   Box Model

Despite taking a simple approach to a complex problem, the box model provided useful insights into how the distribution of dissolved oxygen in the deepwater Gulf might be established and maintained. Foremost is the finding that the distribution of dissolved oxygen is primarily controlled by extra-basin effects, i.e., the source of dissolved oxygen to the deep waters is the transport and mixing of oxygen-rich water masses into the Gulf of Mexico through the Yucatan Channel. Therefore, the success of any additional modeling efforts will be critically dependent on obtaining accurate estimates of the transport and, more importantly, its distribution with depth. Based on numerous model scenarios, the modeled flow through the Yucatan Channel at depth is insufficient to maintain the observed oxygen profiles in the deepwater Gulf. The model suggests that approximately 30% of the Yucatan flow is diverted into the Gulf, although this may be a model artifact. Unfortunately the model is unable to provide an independent estimate of the flux necessary to ventilate the deep Gulf waters below the 2000-m sill depth of the Yucatan Channel. This vertical mass flux is critical to maintaining the deep oxygen profiles.

Within the basin, the model indicates there are differing secondary controls depending on the depth. In the surface layer (0-200 m), the model-determined net production of oxygen is consistent with the rates reported for air-sea exchanges and photosynthesis at the latitude associated with the Gulf of Mexico. In the upper 800 m, the model indicates that the oxidation of carbon in the water column plays a secondary, but important, role to that of transport in maintaining oxygen profiles. This is evidenced by the model-determined oxidation rates that would cause the oxygen concentrations within the Gulf to decay to half of their present value in ten years *if the Gulf was isolated from any transport*. Below 1500 m, the model suggests that the oxidation in the water column is not necessary to maintain oxygen levels, but that the oxidation at the sediment interface plays a minor, but necessary, role.

1.5 Major Results

Spatial and Temporal Changes in Oxygen Levels: Early studies, using known high quality data sets, found that below approximately 1500 m there was "no clearly discernable horizontal variation in dissolved oxygen in these waters" and "only slight vertical oxygen gradients" (Nowlin et al. 1969). Analysis of the additional data sets used in this study suggest the deep Gulf waters may have slightly different oxygen values in three different regions: the southeastern Gulf with mean dissolved oxygen concentrations of 4.99 mL·L^{-1}, the northern Gulf with a mean of 4.95 mL·L^{-1}, and the southwestern Gulf with a mean of 4.87 mL·L^{-1}. However, the quality of the data sets is variable enough that this finding is not definitive.

Comparison of data sets from the 1970s with those from 2000/2001 indicates there has not been any discernible change in the vertical or horizontal distribution of dissolved oxygen in the Gulf of Mexico. This suggests that the transport mechanisms that replenish the oxygen are adequate to balance the oxygen consumption from decay of organic matter, including that from oil seeps and anthropogenic sources.

Influences of Shelf Processes on Deepwater Dissolved Oxygen Concentrations: Data sets available are not adequate to study the details of the effects of shelf-deep ocean exchanges on the oxygen levels. However, they do show no water mass of consequence is formed on the shelves. There are no shelf waters dense enough to sink to depth into the deep ocean of the Gulf and so to provide a source of ventilation for the deep waters. Any shelf-deep ocean exchanges will impact only the upper waters. Transport of organic matter off the shelf, particularly in the region of the Mississippi River Delta, could have local effects on the oxygen concentrations in the deep Gulf if the material sinks to depth before decaying, but evidence of this effect is marginal.

Effects of Natural Hydrocarbons on Dissolved Oxygen Concentrations: NRC (2003) reports an estimated 140,000 ± 60,000 tonnes of hydrocarbons are introduced into the Gulf of Mexico from natural seeps each year. This seepage begins at the sea floor and rises through the water column to the surface, so it potentially can affect the total water column. Although this natural leakage of hydrocarbons into the Gulf has been occurring for millions of years, the values of dissolved oxygen throughout the water column have been stable over the period during which measurements are available and no significant large-scale perturbations in the dissolved oxygen content of the deep Gulf waters have been observed to result from this input. However, localized low oxygen conditions are reported to occur within a few centimeters of the sediments and, hence, it is likely that oxygen concentrations of the waters at the immediate sediment-water interface may be depleted at the seeps or associated brine pools. Chemosynthetic communities are active at the seep sites. Because they require access to oxygenated waters, their presence is additional evidence of ventilation of the deepest Gulf waters.

Effects of Oil and Gas Activities on Dissolved Oxygen Concentrations: NRC (2003) reports anthropogenic sources of hydrocarbons to the Gulf of Mexico are estimated at approximately 2000 tonnes from extraction (1700 tonnes of which are from produced waters), 1600 tonnes from transportation, and 6800 tonnes from consumption. These inputs are widespread throughout the Gulf. Thus, their impact on dissolved oxygen concentrations is negligible over the large scale in the deepwater Gulf of Mexico. The localized effects, however, might be measurable and, in some cases, substantial. Most of these discharges occur at or near the sea surface, so there would be essentially no effect on the dissolved oxygen concentrations of the deep waters. An exception might be any resulting increase in the quantity of decaying organic material, such as from growth of organisms that utilize the hydrocarbons, that sinks through the water column.

Catastrophic oil spills can introduce hydrocarbons at 2-3 times the rate of the natural seeps. The effect of such spill on the dissolved oxygen concentrations would depend on many factors, including chemical characteristics of the spilled material, location in the water column and nature of the spill, residence time in the water column of labile components of the hydrocarbon as the spilled material rises toward the sea surface, and environmental conditions at the spill site, including sites to which the material may be transported. In the large scale, the impact is expected to be minimal. Effects on dissolved oxygen concentrations in the deepwater Gulf from anthropogenic hydrocarbon discharges likely would be most substantial at the sediment-water interface at a discharge site or at the sea surface reached by the plume, not within the water column itself. Effects on dissolved oxygen concentrations are expected to be local, but potentially could be severe for short periods. Further study is needed to examine such effects because they are localized and complex.

Determination of local effects is a priority. However, knowledge of many factors, such as (1) the environmental conditions (e.g., currents at all affected depths, wave action, water temperatures, local biological characteristics of the ecosystems that are exposed) at the discharge location, (2) the type and rate of the discharge, and (3) the chemical nature of the material discharged, is necessary to make a thorough assessment of the effects of various oil and gas activities on dissolved oxygen concentrations in the Gulf of Mexico. This is a highly complex problem that requires a more sophisticated approach that can be accomplished by the simple box model and the data reanalysis in this study.

1.6   Information Gaps and Recommendations

A number of data gaps have been identified. These include: sparse dissolved oxygen data from depths $\geq$ 1500 m; lack of observations on circulation, particularly below 1500 m, to determine horizontal advection and vertical fluxes and associated mechanisms; no measurements of the rate of consumption of dissolved oxygen in the deep waters; lack of data on the age of the water masses; and lack of information on the carbon cycle in the deep water column. These gaps could be addressed by a combination of a high quality, World Ocean Circulation Experiment (WOCE 1991) type of hydrographic survey, including the full suite of tracers, chemical measurements, and shipboard current observations, with measurements of currents (moorings and/or floats) and detailed modeling studies.

A higher priority would be to determine whether anthropogenic activities are *locally* affecting dissolved oxygen levels. This will require local monitoring of dissolved oxygen concentrations at discharge sites. A study designed to assess the nature and extent of the effects on dissolved oxygen concentrations of discharges at selected sites of oil and gas exploration and production operations in the deepwater Gulf should be undertaken. Monitoring of discharges should include measurements to determine the fate of drill cuttings, drilling fluids if discharged, produced waters, and similar discharges. Local environmental conditions, such as currents, waves, temperature, salinity, nutrients, carbon, particulate matter, and other chemical properties should be monitored at the selected locations to allow determination of the importance of the various complex of factors involved and to provide information that would allow development of an oil spill model that could assess possible effects on dissolved oxygen of a hypothetical subsurface blowout. Only if substantive effects on the dissolved oxygen concentrations are shown to occur should detailed modeling, process, or other studies be undertaken.

2 INTRODUCTION

2.1   Background

With the increase of oil and gas activities in the deepwater Gulf of Mexico, the Minerals Management Service (MMS), U. S. Department of the Interior, seeks a greater understanding of the environment in Gulf of Mexico waters deeper than 400 m to aid in making decisions about the protection of the marine environment. The MMS awarded the contract for the "Deepwater Program: Understanding the Processes that Maintain the Oxygen Levels In the Deep Gulf of Mexico" to the Texas A&M Research Foundation in July 2002. Under the contract, scientists at Texas A&M University (TAMU) conducted the study. Because discharges of oil or gas can consume oxygen during degradation, this study considers what is known about the distribution of oxygen and processes that control it, what the information gaps might be, and how those gaps might be filled.

The study area is the region of the Gulf of Mexico basin with water depths of 400 m or more (Figure 2.1). The data sets and analyses used in this study, however, were extended shoreward to the 200-m isobath, which is the approximate location of the shelf edge. The study area contains the slopes off the United States and Mexico, and the Sigsbee and Florida abyssal plains of the Gulf of Mexico basin. It includes the Yucatan Channel and Florida Straits. Waters of interest to the study extend from the sea surface to the bottom, which can reach over 3700 m.

2.2   Study Objectives and Description

The objective of this study is to understand the types and rates of processes occurring in the deep Gulf of Mexico that affect the levels of oxygen in the deepwater and the balance that maintains this level. The objective will be addressed by four sub-objectives:

(1)   conducting a data search, synthesis, and reanalysis of available historical oxygen data collected from water depths greater than 400 meters throughout the entire Gulf of Mexico basin;

(2)   developing a simple box model that describes the sources and sinks of oxygen in the deep Gulf and that can be used to evaluate inputs from oil and gas activities, using parameters and constants determined from item 1;

(3)   identifying data gaps including an evaluation of the quality of the data and how representative it is of the Gulf's basin environment; and

(4)   proposing methods for filling the data gaps which could include field work.

To accomplish these objectives, three tasks were set out for the Oxygen Study. Each task had a Principal Investigator (PI) and one or more Co-PIs (Table 2.1). They were responsible directly to the Program Manager, Ann E. Jochens, for successful completion of their task. All members participated in the synthesis and interpretation work and in preparation of this synthesis report. Matthew K. Howard was the Data Manager, who provided data administration and data archival per the terms of the contract.

The three tasks were:

Task 1. Data Search, Synthesis, and Reanalysis: Historical dissolved oxygen data that have been collected within the Gulf of Mexico basin were assembled and synthesized for this report. Along with the data, the methodology for collecting the data was identified where possible. All data

5

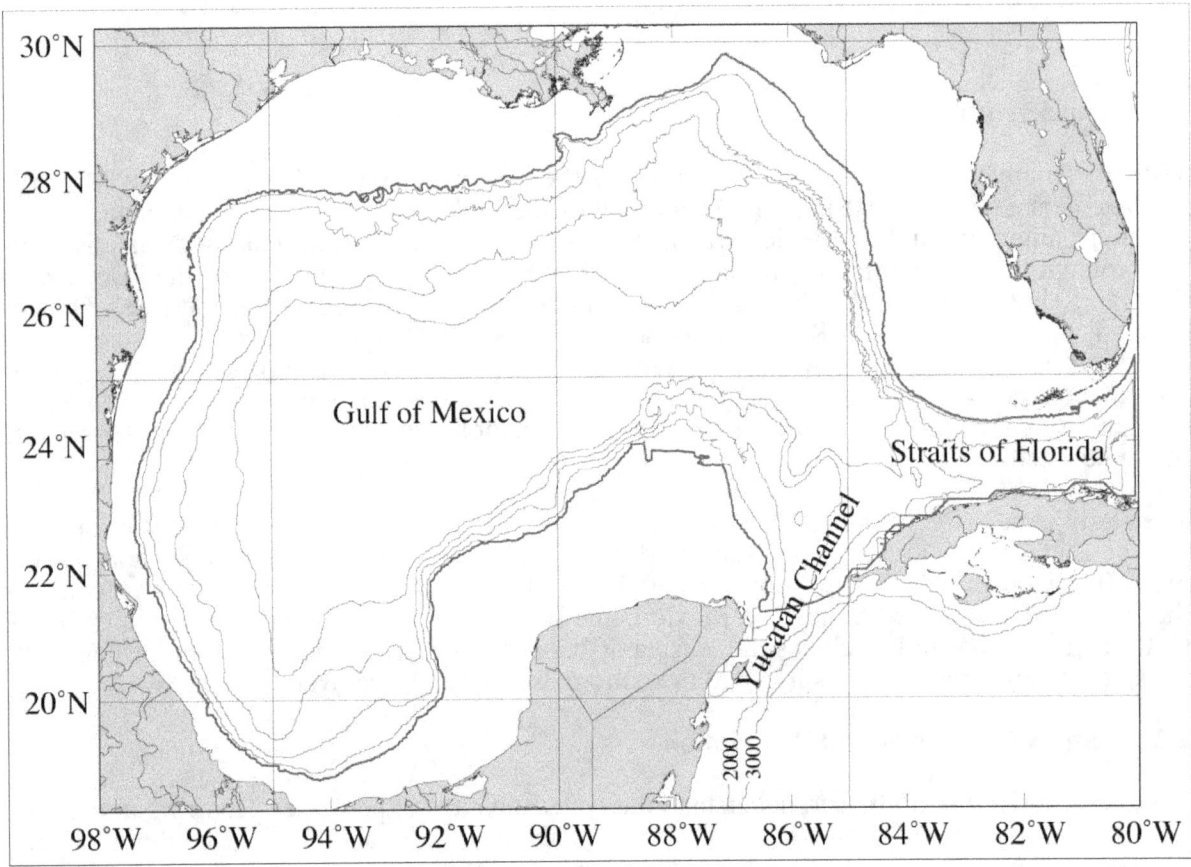

Figure 2.1.   Study area for the reanalysis of the historical oxygen data. The study area is
                within the boundary denoted by the blue line. This boundary is the 200-m isobath
                in the Gulf and the entrances of the Yucatan Channel and Florida Straits.
                Bathymetric contours are 200, 400, 1000, 2000, and 3000 m.

were checked for quality, including assessment of sample collection and analysis methodology, and flagged if determined to be in question. Trends in the data, such as differences between eastern and western basins, a statistical analysis of the data, and flushing rates of the Gulf of Mexico were evaluated (see Section 4.3 below).

Task 2. Development of a Simple Box Model: A simple box model was constructed to represent the inputs to and outputs from the deep Gulf of Mexico basin. It was developed to address, as well as such a model can, the complexity of the distribution of oxygen. It included the oxygen minimum zone. An early version considered the eastern and western basins of the Gulf of Mexico, but this was found to be too complex a system given the data sets available (see Section 5). East-west differences, however, were addressed as part of the data reanalysis work under Task 1 (Section 4.3). The model was used to help estimate the flushing time of the Gulf of Mexico and the potential impacts from oil and gas activities on dissolved oxygen in the water column and in surficial sediments. These issues also were addressed by additional calculations (see Sections 4.1.3 and 4.4).

Table 2.1

Scientific Personnel for the Oxygen Study
(The Principal Investigator (PI) is identified for each task.)

Description	Personnel
Program Management	Ann E. Jochens, Program Manager
Data and Information Management	Matthew K. Howard, Data Manager
Interpretation and Synthesis Team	Ann Jochens, Team Leader
	Les Bender
	Steve DiMarco
	Matt Howard
	Mahlon Kennicutt
	John Morse
	Worth Nowlin
Task 1 (Data Assembly and QA/QC)	Ann Jochens, PI
	Steve DiMarco, Co-PI
	Matt Howard, Data Manager
	Worth Nowlin
	assistance from Task 2 team
Task 2 (Box Model)	Les Bender, PI
	Mahlon Kennicutt, Co-PI
	John Morse, Co-PI
	assistance from Task 1 team
Task 3 (Gaps Identification)	Ann Jochens, PI
	Mahlon Kennicutt, Co-PI
	Task 1 members
	Task 2 members

Task 3. Identify Information Gaps: From the box model and data synthesis, information gaps were identified. These included identification of gaps as seen by the inventory of data and metadata and of gaps in knowledge related to physical and/or biogeochemical processes and biogeochemical rates from evaluation of the box model results. Methods for filling the gaps, such as field work, modeling, or a combination, were developed. Resulting recommendations are described in Section 6 of this report.

2.3 Science Review Group

The Science Review Group (SRG) for the Oxygen Study is composed of three members from the oceanographic community. Table 2.2 shows the members and their affiliations. The purpose of the SRG is to review the progress and scientific value of the study, recommend improvements, and review and comment on the draft Synthesis Report. Quarterly Reports including a Technical Summary were provided to the SRG for review and comment during the Study.

Table 2.2

Members of the Oxygen Study Science Review Group

Member	Affiliation
Dr. Larry P. Atkinson	Old Dominion University
Dr. Robert M. Key	Princeton University
Dr. John M. Morrison	North Carolina State University

2.4   Report Organization

This report presents the synthesis of the data assembled and the scientific results of the Oxygen Study. The Executive Summary, Section 1, provides a brief review of the findings, results, and significance of this study. Descriptions of the data base assembled and the quality control processing methods used are given in Section 3. Background information on the physical oceanography and water masses of the Gulf of Mexico, description of physical and biogeochemical sources and sinks of dissolved oxygen, and the results of the data reanalysis are presented in Section 4. The simple box model formulated to aid in the interpretation of the data distributions is given in Section 5. Information gaps and recommendations for future studies are presented in Section 6. References are provided in Section 7.

3 DATA SETS

The objective of this study is to understand the types and rates of processes occurring in the deep Gulf of Mexico that affect the levels of oxygen in the deepwater and the balance that maintains this level. The objective was met first by conducting a data search, synthesis, and reanalysis of all oxygen data collected from water depths greater than 200 meters throughout the entire Gulf of Mexico basin. This chapter presents the results of the data search, assembly, and quality control processing.

3.1   Data Assembly

Dissolved oxygen data sets in the deepwater Gulf of Mexico have been identified through search of the Texas A&M University (TAMU) Deepwater Physical Oceanography data archive and review of the literature that identifies cruises on which oxygen data were collected. The TAMU Deepwater Physical Oceanography data archive was initially developed for the MMS-funded Deepwater Reanalysis and Synthesis of Historical Data Study (Deepwater; Nowlin et al. 2001). This archive contains the then available physical oceanographic data, including dissolved oxygen concentrations, from the Gulf of Mexico in water column depths greater than 200 m (DiMarco et al. 2001). Data were acquired from MMS, other state and federal agencies, national laboratories, universities, Mexican institutions, and the private sector, especially the oil and gas industry. The TAMU Deepwater Physical Oceanographic archive was searched, and stations with oxygen data were extracted into a separate database (Oxygen Archive) for this study.

A brief survey was made of literature that contained discussion of dissolved oxygen in the deepwater Gulf of Mexico. The review included peer-reviewed journals as well as readily available technical and data reports, manuscripts, and other documents. The results of the literature review were cross-matched with the Deepwater inventory to assess whether referenced data sets were accounted for in the archive. Data sets identified as not being in the archive were sought by contacting either the individual or institution responsible for their collection and/or by contacting authors of publications or reports in which the data were discussed. Data obtained were added to the Oxygen Archive. Some missing data sets were deemed as lost or inaccessible.

The temporal range of hydrographic data (which includes temperature, salinity, and dissolved oxygen observations) currently in the Oxygen Archive covers the range from 1922 through 2001. The data sources from which this database was assembled are:

•   CD-ROM NODC-02 "Global Ocean Temperature and Salinity Profiles Vol. 1 - Atlantic, Indian, and Polar Oceans." U.S. Dept. of Commerce. NOAA/NESDIS/NODC. 1991. This CD-ROM contains data collected by various researchers and institutions between 1900 and 1990 and includes oceanographic station (i.e., bottle) data and CTD/STD data.

•   CD-ROM NODC-72 "Gulf of Mexico Hydrographic Data. CTD, XBT, and Bottle Data 1987-1995 and GulfCET Program CTD, XBT, and Marine Mammals Database, 1992-1994." December 1996. U.S. Dept. of Commerce, NODC. A collection of station data from the Texas A&M University Ship of Opportunity (TIGER/SOOP) and MMS GulfCET programs.

•   CD-ROM NODC-126 to NODC-130 "WOD98 World Ocean Data Base 1998 Version 2." Jan 2000. U.S. Dept. of Commerce. NOAA/ NESDIS/NODC/OCL. Data sets includes bottle and CTD data from 1900 to 1995. This is a revised version of World Ocean Atlas 1994. Some duplicate stations were removed and many new stations added from data archaeology and data rescue programs. A 5-CD-ROM set.

9

•   CD-ROM NODC-92 "Texas-Louisiana Shelf Circulation and Transport Processes Study. Hydrography, Drifters, ADCP, and Miscellaneous Sensors Data and Reports 1992-1994." September 1998. U.S. Minerals Management Service and U.S. Dept. of Commerce. NOAA/NODC. A 5-CD-ROM set. This archive contains bottle and CTD data. Most data are for the waters of the continental shelf but some samples were taken in water as deep as 1000 m.

•   CD-ROM NODC 148 "World Ocean Database 2001, Ocean Station and Surface Data, Observed Depth Data" – an updated version of the WOD98. This is part of an 8-CD-ROM set prepared by the Ocean Climate Laboratory for the National Oceanographic Data Center, Silver Spring, MD 20910. Data are also available online through various NOAA portals.

•   CD-ROM "Northeastern Gulf of Mexico Physical Oceanography Program. Chemical Oceanography and Hydrography Study. Data and Reports." 2002. Dept. of Oceanography, Texas A&M University and U.S. Dept. of the Interior. Minerals Management Service. This contains hydrographic data, including dissolved oxygen data, collected from the northeastern Gulf shelves out to 1000-m water depth between 1997 and 2000.

•   CD-ROM "Deepwater Physical Oceanography Reanalysis and Synthesis of Historical Data. Deepwater CD-ROM 2 Hydrography and expendable instruments data." April 2001. U.S. Dept. of the Interior. Minerals Management Service. A 3-CD-ROM set. This contains available physical oceanographic data for the deep (greater than 200 m) Gulf of Mexico. These data were assembled from the above CD-ROM sources and from previously unreleased, non-proprietary oil industry surveys.

•   Originator data files from the cruises in May/June 2000 and June 2001 aboard the R/V *Gyre*, conducted as part of the MMS-funded "Deepwater Program: Northern Gulf of Mexico Continental Slope Habitats and Benthic Ecology" Study (DGOMB). Oxygen data were collected on the June 2002 DGOMB cruise, but were discarded by the originator because the analysis instrumentation was faulty and all analyses were bad.

•   Data sets typed in and checked from several TAMU Oceanography Technical Reports and TAMU Data Reports for various cruises of the R/V *Jakkula*, *Hidalgo*, *Alaminos*, and *Gyre*. Data from these cruises were unavailable through NODC and were not digitally available.

•   Some data were found in NODC's Ocean Archive System. This online resource serves data sets in the originator's format together with the originator's metadata. Sixty-one data sets spanning 1969-1995 were downloaded and scanned for oxygen data. 1339 profiles were extracted, reformatted and examined. Of these 230 were from deep Gulf stations, the rest were from the shelf or just outside the study region.

•   National Marine Fisheries Service (NMFS): The Southeast Area Monitoring and Assessment Program (SEAMAP) is a cooperative program between NMFS and the five Gulf States fishery management agencies. Mark Mcduff, a data manager for the program, kindly provided the entire SEAMAP environmental database containing 16383 profiles. Most were from stations over the shelf but 319 stations occurred in the deepwater study area.

•   Various individuals who kindly provided data sets missing from the NODC archive.

The number of stations with discrete oxygen data is approximately 3600 from over 200 cruises. There are nearly 35,000 discrete data points that were collected between 1922 and 2003. Additionally, there are over 1800 CTD casts with oxygen sensor data. Table 3.1 gives the cruises on which bottle oxygen data were identified as having been collected, information on the cruise dates, an indicator of whether the data are in the database, and any references consulted for the particular data set. Table 3.2 gives the cruises on which oxygen sensor data were collected.

10

Table 3.1

Cruises with Bottle Oxygen Data
(The data sources are the data originator, if known, or the data repository, usually NODC, from which the data were obtained. Dates give the range of sampling dates. The number of stations in the deepwater data base are given. References to literature and/or reports are shown if known.)

Vessel	Cruise ID	Dates (mm/dd/yyyy)	No. Stns	Data Source	References
Dana II*		02/01/1922-02/06/1922	10	CCF	
Atlantis I*		05/04/1933-05/05/1933	5	WHOI	Wennekens 1959
Atlantis I*		03/01/1934-03/041934	10	WHOI	
Atlantis I*		02/16/1935-04/13/1935	66	WHOI	Wennekens 1959; Le Bureau Du Conseil Service Hydrographique 1936; Nowlin et al. 1969; Nowlin 1972
San Pablo*	7	12/02/1950-12/09/1950	14	NOO	
Rehoboth*	7	12/03/1950-12/09/1950	13	NOO	
A. A. Jakkula	54-09	07/08/1954-07/17/1954	35	TAMU	Leipper 1956
A. A. Jakkula	54-10	08/14/1954-09/07/1954	40	TAMU	Leipper 1956; Austin 1955; Wennekens 1959
A. A. Jakkula	55-1	01/12/1955-02/27/1955	53	TAMU	Drummond 1956
A. A. Jakkula	55-3	03/29/1955-04/07/1955	13	TAMU	Drummond 1956
A. A. Jakkula	55-8	07/04/1955-07/06/1955	6	TAMU	Drummond 1957
Caryn*	96	05/17/1956-05/18/1956	7	WHOI	Wennekens 1959
Unknown (U of M.)	G5705	04/01/1957-04/07/1957	0		Wennekens 1959; 7 stations, no data found
Gerda*		10/20/1957	5	NODC	
Gerda*		02/22/1958-02/24/1958	6	NODC	
Hidalgo*	58-H-1	03/27/1958-04/01/1958	10	TAMU	McLellan 1959; McLellan 1960; Nowlin et al. 1969; Nowlin 1972
Hidalgo	58-H-4	05/14/1958-06/28/1958	20	TAMU	McLellan 1959; McLellan 1960; Nowlin et al. 1969; Nowlin 1972
Gerda*		04/11/1958-04/25/1958	10	NODC	
Gerda*		09/30/1958-10-01/1958	11	NODC	
Hidalgo*	59-H-2	02/18/1959-02/26/1959	10	TAMU	McLellan 1960; Nowlin et al. 1969; Nowlin 1972
Hidalgo*	60-H-6	05/19/1960-05/28/1960	15	TAMU	
Gerda*	6110	03/01/1961-03/04/1961	10	MIAMI	
Gerda	6110	04/30/1961	1	NODC	
Gerda*	6110	11/18/1961-11/20/1961	4	NODC	
Hidalgo*	61-H-6	03/23/1961-04/07/1961	6	TAMU	
Hidalgo*	61-H-9	05/23/1961-06/06/1961	16	TAMU	
Hidalgo*	61-H-16	10/15/1961-10/17/1961	5	TAMU	
Hidalgo*	61-H-19	11/11/1961-11/13/1961	5	TAMU	
Hidalgo*	62-H-1	01/17/1962-01/19/1962	5	TAMU	
Hidalgo*	62-H-3	02/14/1962-03/31/1962	90	TAMU	McLellan & Nowlin 1962; McLellan & Nowlin 1963; Nowlin & McLellan 1967; Nowlin et al. 1969; Nowlin 1972; Ichiye et al. 1973
Gerda*		04/24/1962-04/25/1962	3	NODC	
Hidalgo*	62-H-4	05/12/1962-05/25/1962	10	TAMU	
Gerda*		06/07/1962-06/14/1962	14	NODC	
Alaminos*		12/07/1963-12/08/1963	4	NODC	
Alaminos*	64-A-2	01/17/1964-01/25/1964	17	TAMU	Nowlin et al. 1969; Nowlin 1972
Olenty*		02/24/1964-03/17/1964	9	NODC	
Alaminos*	64-A-3	02/27/1964-03/02/1964	7	TAMU	Nowlin et al. 1969; Nowlin 1972
Obdorsk*	SRT-R9029	05/23/1964-06/14/1964	12	ARI	
Alaminos*	65-A-1	02/07/1965	4	TAMU	Leipper 1968b
Akademik A. Kovalevsky*		03/28/1965-05/31/1965	45	ARI	
Alaminos*	65-A-11	08/11/1965-08/21/1965	23	TAMU	Leipper 1968a
Alaminos*	65-A-13	09/12/1965-09/23/1965	59	TAMU	Leipper 1968a; Ichiye et al. 1973
Alaminos*		10/02/1965-10/07/1965	6	NODC	
Geronimo*	6	11/01/1965-11/02/1965	6	NMFS-M	
Alaminos*		01/09/1966-01/27/1966	9	NODC	
Pillsbury*		01/31/1966	1	MIAMI	
Alaminos*	66-A-3	02/11/1966-02/20/1966	42	TAMU	Leipper 1968b
Undaunted*		06/02/1966-06/06/1966	12	NODC	

Table 3.1

Cruises with Bottle Oxygen Data (continued)

Vessel	Cruise ID	Dates (mm/dd/yyyy)	No. Stns	Data Source	References
*Alaminos**	66-A-8	06/06/1966-06/22/1966	14	TAMU	Nowlin et al. 1969; Nowlin 1972; Leipper 1970
*J. E. Pillsbury**	6609	06/12/1966-06/22/1966	2	MIAMI	
*SRT-R9029**	CUBA I	08/12/1966	3	FRC	
Alaminos	66-A-11	08/05/1966-08/07/1966	10	TAMU	Leipper 1968a; Ichiye et al. 1973
*Alaminos**	66-A-15	11/06/1966-11/12/1966	9	TAMU	Leipper 1968a; Ichiye et al. 1973
Alaminos	66-A-15	10/28/1966-11/04/1966	29	TAMU	Leipper 1968a; Ichiye et al. 1973
*SRT-R9029**	CUBA III	01/22/1967-01/23/1967	3	FRC	
*Peto**	CUBA III	02/18/1967	4	FRC	
*SRT-R9029**	CUBA III	05/06/1967	3	FRC	
*Mikhail Lomonosov**	20	02/21/1967	1	ARI	
*Mikhail Lomonosov**	20	03/02/1967	1	ARI	
*Alaminos**	67-A-4	06/11/1967-07/01/1967	24	TAMU	Nowlin et al. 1969; Nowlin 1972; Ichiye et al. 1973
*Alaminos**	67-A-6	08/05/1967-08/20/1967	28	TAMU	Leipper 1968a
*SRT-R9074**	CUBA II	08/26/1967	4	FRC	
*SRT-R9074**	CUBA II	09/23/1967-09/24/1967	3	FRC	
*SRT-R9074**	CUBA II	10/23/1967	3	FRC	
*Alaminos**	67-A-8	09/09/1967-09/12/1967	8	TAMU	Nowlin et al. 1969; Nowlin 1972
*SRT-R9074**	CUBA IV	01/15/1968-01/16/1968	3	FRC	
*SRT-R9074**	CUBA IV	02/02/1968-02/03/1968	3	FRC	
*SRT-R9074**	CUBA IV	04/11/1968-04/12/1968	3	FRC	
*Alaminos**	68-A-2	02/16/1968-03/05/1968	26	TAMU	Leipper 1968b
*J. E. Pillsbury**	P-6803	04/11/1968-04/27/1968	13	MIAMI	
*Alaminos**	68-A-5	04/22/1968-05/13/1968	16	SUSIO	
*Sardina**	CUBA IV	06/29/1968	4	FRC	
*Alaminos**	68-A-8	08/18/1968-09/03/1968	21	TAMU	Leipper 1968c
*Alaminos**	68-A-9	09/11/1968-09/15/1968	8	TAMU	
*SRT-R9029**	CUBA IV	09/12/1968	3	FRC	
*SRT-R9029**	CUBA IV	10/15/1968-10/16/1968	4	FRC	
*Atlantis II**		11/16/1968-11/25/1968	3	BLOS/ MIAMI	
*SRT-R9112**	CUBA V	02/13/1969	3	FRC	
*SRT-R9112**	CUBA V	04/06/1969-04/07/1969	3	FRC	
*SRT-R9112**	CUBA V	05/25/1969-05/26/1969	2	FRC	
*Trident**	68	04/19/1969-04/24/1969	8	URI	
*Alaminos**	69-A-7	05/01/1969-05/18/1969	14	SUSIO	
*Alaminos**	69-A-8	06/08/1969-06/15/1969	10	TAMU	
*Kane**	939014	06/10/1969-09/09/1969	148	NOO	
*Alaminos**	69-A-10	07/12/1969-07/29/1969	21	TAMU	
*Akademik N. Knipovich**	CUBA V	07/13/1969-07/14/1969	4	FRC	
*SRT-R9075**	CUBA V	08/02/1969-08/03/1969	2	FRC	
*SRT-R9075**	CUBA V	09/06/1969-09/07/1969	3	FRC	
*SRT-R9075**	CUBA V	10/14/1969	3	FRC	
*Alaminos**	69-A-12	09/09/1969-09/18/1969	6	SUSIO	
*Alaminos**	70-A-3	02/05/1970-03/03/1970	4	SUSIO	
*Alaminos**	70-A-6	04/04/1970-04/21/1970	61	TAMU	
*Alaska**		05/01/1970-05/11/1970	8	TAMU	
*Tursiops**	T7015	05/02/1970-05/11/1970	8	FSU	
*Alaminos**	70-A-7	05/05/1970-05/11/1970	12	SUSIO	
*Alaminos**	70-A-9	06/17/1970-06/21/1970	9	SUSIO	
*Alaminos**	70-A-14	10/22/1970-11/01/1970	20	SUSIO	
*Virgilio Uribe**	C12 (7012)	10/31/1970-11/13/1970	39	NODC	
Virgilio Uribe	VU7012	10/31/1970-11/13/1970	13	NODC	
*Virgilio Uribe**	VU7102	01/18/1971-01/21/1971	6	NIO	
*Aliot (SRT-M8005)**	6189	03/15/1971	1	ARI	
*Aliot (SRT-M8005)**	3234	03/20/1971-03/21/1971	7	CITI	
*Aliot (SRT-M8005)**	3237	07/18/1971	5	CITI	
*Virgilio Uribe**	VU7108	04/28/1971-05/09/1971	19	NIO	
Virgilio Uribe	VU7108	04/28/1971-05/10/1971	31	NIO	
*Virgilio Uribe**	711	05/24/1971-06/09/1971	39	UN	
*Virgilio Uribe**	VU7114	07/21/1971-07/27/1971	19	NIO	
*Virgilio Uribe**	712	08/04/1971-09/03/1971	42	UN	
*Foton(SRT-M8024)**	3314	10/05/1971-10/07/1971	16	CITI	
*Foton**		11/04/1971-11/07/1971	10	CITI	

12

Table 3.1

Cruises with Bottle Oxygen Data (continued)

Vessel	Cruise ID	Dates (mm/dd/yyyy)	No. Stns	Data Source	References
*Foton(SRT-M8024)**	3315	11/04/1971-11/05/1971	6	CITI	
*Foton(SRT-M8024)**	3238	10/06/1971-10/07/1971	3	CITI	
*Foton(SRT-M8024)**	3239	11/06/1971	3	CITI	
*Virgilio Uribe**	VU7120	10/10/1971-10/17/1971	20	NIO	
*Virgilio Uribe**	713	10/27/1971-11/10/1971	38	UN	
*Virgilio Uribe**	VU7202	01/11/1972-01/19/1972	21	NIO	
Virgilio Uribe	VU7202	01/10/1972-01/19/1972	17	NIO	
*Antares**		02/18/1972-02/24/1972	9	NODC	
*Alaminos**	72-A-7	03/26/1972-03/28/1972	2	TAMU	Morrison & Nowlin 1977
*Virgilio Uribe**	721	04/25/1972-05/19/1972	38	UN	
*Alaminos**	72-A-9	05/02/1972-05/21/1972	57	TAMU	Morrison et al. 1973; Morrison & Nowlin 1977
*Tursiops**	11	05/16/1972-05/18/1972	7	FSU	
*Andrey Vilkitsky**		03/17/1973	1		
*Antares**		03/23/1973-03/24/1973	3		
*Akademik Kurchatov**	14	04/03/1973-04/07/1973	4	IO	
*Alaminos**	73-A-8	05/18/1973-06/03/1973	14	SUSIO	
*Virgilio Uribe**	VU7310	05/23/1973-06/05/1973	27	UN	
*Researcher**		10/06/1976-11/14/1976	89	AOML	Berberian and Cantillo 1976
*Gyre**	76-G-11	11/11/1976-11/16/1976	2	TAMU	Key 1981
Gyre	77-G-14	12/05/1977-12/06/1977	2	TAMU	
Gyre	78-G-03	04/04/1978-04/12/1978	42	TAMU	Morrison et al. 1983
*Gyre**		04/06/1982	1		
*Suncoaster**		09/18/1982	1		
*Suncoaster**	GM01	03/09/1983-03/20/1983	35	SAIC	SAIC 1986
*Suncoaster**	SC8310	11/12/1983-11/18/1983	41	SAIC	SAIC 1986
*Cape Florida**	CF8405	05/06/1984-05/17/1984	39	SAIC	SAIC 1986; SAIC 1987
*Pelican**	PN-8502	10/22/1985-10/25/1985	16	SAIC	SAIC 1988
*Justo Sierra**	JS085	03/19/1985	1		
*Altair**	AL8601	01/24/1986-02/04/1986	45	SAIC	SAIC 1988
*Pelican**	PN8713	04/08/1987	2	SAIC	SAIC 1989
Pelican	PN8713	04/07/1987-04/10/1987	22	SAIC	SAIC 1989
Pelican	PN8715	04/25/1987-04/26/1987	15	SAIC	SAIC 1989
Pelican	PN8803	07/17/1987	11	SAIC	SAIC 1989
Pelican	PN8814	11/15/1987-11/16/1987	9	SAIC	SAIC 1989
*Gyre**	87G04	04/12/1987-04/16/1987	11	TAMU	
*Gyre**	87G10	10/27/1987-10/28/1987	3	TAMU	
*Gyre**	87G11	11/18/1987-11/23/1987	12	TAMU	
*Gyre**	87G12	11/29/1987- 12/05/1987	11	TAMU	
*Gyre**	88G05	10/171988-10/23/1988	19	TAMU	
Pelican	PN8820	02/09/1988-02/10/1988	15	SAIC	SAIC 1989
Pelican	PN8903	07/21/1988	16	SAIC	SAIC 1989
*Gyre**	89G06	05/18/1989-05/201989	4	TAMU	
*Gyre**	89G15	11/13/1989-1117/1989	10	TAMU	Biggs 1989
*Gyre**	90G04	02/21/1990	1	TAMU	
*Gyre**	90G05	02/28/1990	1	TAMU	
*Gyre**	90G10	07/14/1990-07/23/1990	12	TAMU	Biggs 1990
*Gyre**	90G15	07/13/1990-07/15/1990	6	TAMU	Biggs 1991a
*Gyre**		10/13/1990-10/15/1990	7	TAMU	
*Gyre**	91G02	03/05/1991-03/09/1991	8	TAMU	Biggs 1991b
*Gyre**	91G04	06/15/1991-06/25/1991	11	TAMU	Biggs 1991c
*Gyre**	92G04	04/08/1992	1	TAMU	
*Gyre**	92G05	05/01/1992-05/08/1992	7	TAMU	LATEX: Jochens et al. 1998; Nowlin et al. 1998a, 1998b
*Gyre**	92G07	06/22/1992-06/25/1992	3	TAMU	
*Gyre**	92G08	08/01/1992-/0807/1992	12	TAMU	LATEX: Jochens et al. 1998; Nowlin et al. 1998a, 1998b
*Gyre**	92G10	10/05/1992	2	TAMU	
*J. W. Powell**	92P10	11/05/1992-11/12/1992	8	TAMU	LATEX: Jochens et al. 1998; Nowlin et al. 1998a, 1998b
*Gyre**	93G01	01/08/1993-01/14/1993	10	TAMU	
*Gyre**	93G02	02/06/1993-02/12/1993	12	TAMU	LATEX: Jochens et al. 1998; Nowlin et al. 1998a, 1998b
*J. W. Powell**	93P06	04/28/1993-05/10/1993	21	TAMU	LATEX: Jochens et al. 1998; Nowlin et al. 1998a, 1998b
*J. W. Powell**	93P11	07/28/1993-08/06/1993	21	TAMU	LATEX: Jochens et al. 1998; Nowlin et al. 1998a, 1998b
*J. W. Powell**	93P14	11/10/1993-11/21/1993	23	TAMU	LATEX: Jochens et al. 1998; Nowlin et al. 1998a, 1998b

Table 3.1

Cruises with Bottle Oxygen Data (continued)

Vessel	Cruise ID	Dates (mm/dd/yyyy)	No. Stns	Data Source	References
*Gyre**	94G01	04/25/1994-05/07/1994	40	TAMU	LATEX: Jochens et al. 1998; Nowlin et al. 1998a, 1998b
*Gyre**	94G03	05/17/1994	1	TAMU	Biggs 1994a
*Gyre**	94G05	07/19/1994-07/20/1994	8	TAMU	Biggs 1994b
*Gyre**	94G07	08/16/1994-08/19/1994	4	TAMU	Biggs 1994c
*Gyre**	94G08	10/20/1994-10/24/1994	33	TAMU	Biggs 1995a
*J. W. Powell**	94P10	07/28/1994-08/07/1994	41	TAMU	LATEX: Jochens et al. 1998; Nowlin et al. 1998a, 1998b
*Gyre**	94G09	11/03/1994-11/13/1994	49	TAMU	LATEX: Jochens et al. 1998; Nowlin et al. 1998a, 1998b
*Gyre**	95G03	06/13/1995-06/16/1995	7	TAMU	Biggs 1995b
*Gyre**	97G14	11/17/1997-11/26/1997	47	TAMU	NEGOM: Jochens et al. 2002
*Gyre**	98G05	05/06/1998-05/15/1998	47	TAMU	NEGOM: Jochens et al. 2002
*Gyre**	98G10	07/27/1998-08/06/1998	42	TAMU	NEGOM: Jochens et al. 2002
*Gyre**	98G15	11/14/1998-11/24/1998	45	TAMU	NEGOM: Jochens et al. 2002
*Gyre**	99G07	05/17/1999-05/27/1999	43	TAMU	NEGOM: Jochens et al. 2002
*Gyre**	99G08	08/18/1999-08/27/1999	43	TAMU	NEGOM: Jochens et al. 2002
*Gyre**	99G12	11/13/1999-11/21/1999	44	TAMU	NEGOM: Jochens et al. 2002
*Gyre**	00G04	04/16/2000-04/26/2000	48	TAMU	NEGOM: Jochens et al. 2002
*Gyre**	00G05	05/04/2000-06/17/2000	44	TAMU	DGoMB: Rowe 2005
*Gyre**	00G08	07/29/2000-08/07/2000	46	TAMU	NEGOM: Jochens et al. 2002
*Gyre**	01G05	06/03/2001-06/18/2001	12	TAMU	DGoMB: Rowe 2005
Various	SEAMAP	06/02/1982-09/23/2003	319	SEAMAP	http://www.gsmfc.org/seamap.html

* denotes cruise data were used in statistical analyses; for cruises listed twice, one listing gives the number of stations used in statistical analyses and the other is the number not used

Key to Data Sources:
AOML: NOAA, Atlantic Oceanographic and Meteorological Laboratories
ARI: Atlantic Research Institute of Fishing Econ. and Ocean. (ATLATNIRO), USSR
BLOS: Bigelow Laboratory for Ocean Science, ME
CCF: Collection, Carlsberg Foundation, Charlottenlund, Denmark
CITI: Cuban Institute of Technological Investigations, Havana, Cuba
DUKE: Duke University
FRC: Fisheries Research Center, Havana, Cuba
FSU: Florida State University
IO: Institute of Oceanology, AS USSR, Moscow
MHI: Marine Hydrophysical Institution, USSR
MIAMI: Rosenstiel School of Marine and Atmospheric Science, University of Miami
NIO: National Institute of Fisheries, Mexico
NMFS-M: NOAA, National Marine Fisheries Service, Miami, FL
NMFS-G: NOAA, National Marine Fisheries Service, Galveston, TX
NODC: NOAA, National Oceanographic Data Center
NOO: U.S. Naval Oceanographic Office, Bay St. Louis, MS
NOS: NOAA, National Ocean Service, Norfolk, VA
POI: Pacific Oceanological Institute (VLADIVOSTOK)
SAIC: Science Applications International Corporation
SEAMAP: Southeast Area Monitoring and Assessment Program
SIO: Scripps Institute of Oceanography
SUSIO: State University System of Florida, Institute of Oceanography, St. Petersburg, FL
TAMU: Texas A&M University
UN: University National of Mexico, Institute of Geophysics, Mexico City
URI: University of Rhode Island
USCG: U.S. Coast Guard
WCC: Woodward-Clyde Consultants
WDCB: World Data Center B
WHOI: Woods Hole Oceanographic Institution

Table 3.2

Cruises with Oxygen Sensor Data
(The data sources are the data originator, if known, or the data repository, usually NODC, from which the data were obtained.)

Vessel	Cruise ID	Dates (mm/dd/yyyy)	No. Stns	Data Source*
Justo Sierra	JS085	03/05/1985-03/23/1985	27	MMS_DEEPW
OREGON II	CRU_199	04/22/1992-05/23/1992	88	MMS_DEEPW
R/V Gyre	92G05	05/01/1992-05/08/1992	13	MMS_LATEXA
R/V Gyre	92G08	08/01/1992-08/07/1992	19	MMS_LATEXA
Pelican	CRUISE2	08/23/1992-08/12/1992	31	MMS_DEEPW
R/V Gyre	92G10	11/05/1992-11/12/1992	15	MMS_LATEXA
Pelican	CRUISE3	11/10/1992-11/21/1992	28	MMS_DEEPW
OREGON_II	CRU_203	01/06/1993-02/11/1993	63	MMS_DEEPW
R/V Gyre	93G02	02/06/1993-02/12/1993	19	MMS_LATEXA
Pelican	CRUISE4	02/13/1993-02/24/1993	30	MMS_DEEPW
J.W. Powell	93P06	04/28/1993-05/11/1993	39	MMS_LATEXA
OREGON_II	CRU_204	05/06/1993-06/15/1993	105	MMS_DEEPW
Pelican	CRUISE5	05/25/1993-06/03/1993	30	MMS_DEEPW
J.W. Powell	93P11	06/28/1993-08/06/1993	37	MMS_LATEXA
Pelican	CRUISE6	08/28/1993-09/05/1993	27	MMS_DEEPW
J.W. Powell	93P14	11/10/1993-11/21/1993	37	MMS_LATEXA
Pelican	CRUISE7	12/05/1993-12/12/1993	23	MMS_DEEPW
J.W. Powell	94P10	07/27/1994-08/07/1994	52	MMS_LATEXA
R/V Gyre	94G09	11/02/1994-11/13/1994	51	MMS_LATEXA
SHOYOMARU	1991-01	05/08/1994-05/13/1994	10	MMS_DEEPW
OREGON_II	CRU_209	04/16/1994-06/09/1994	78	MMS_DEEPW
R/V Gyre	96G06	10/12/1996-10/27/1996	12	MMS_DEEPW
Justo Sierra	CANEK_0	12/11/1996-12/18/1996	35	DEEPSTAR
Justo Sierra	CANEK_1	05/25/1997-06/04/1997	52	DEEPSTAR
R/V Gyre	97G08	08/07/1997-08/21/1997	11	MMS_DEEPW
R/V Gyre	97G14	11/16/1997-11/26/1997	48	MMS_NEGOM
R/V Gyre	98G05	11/16/1997-11/26/1998	48	MMS_NEGOM
Justo Sierra	CANEK_2	03/30/1998-04/06/1998	61	DEEPSTAR
R/V Gyre	98G10	07/26/1998-08/06/1998	43	MMS_NEGOM
R/V Gyre	98G15	11/13/1998-11/24/1998	46	MMS_NEGOM
Justo Sierra	CANEK_3	01/31/1999-02/03/1999	28	DEEPSTAR
R/V Gyre	94G01	04/24/1999-05/07/1994	50	MMS_LATEXA
R/V Gyre	99G07	05/20/1999-05/27/1999	46	MMS_NEGOM
R/V Gyre	99G08	08/18/1999-08/27/1999	44	MMS_NEGOM
Justo Sierra	CANEK_4	09/04/1999-09/11/1999	89	DEEPSTAR
R/V Gyre	99G12	11/13/1999-11/21/1999	45	MMS_NEGOM
R/V Gyre	00G04	04/16/2000-04/26/2000	49	MMS_NEGOM
R/V Gyre	00G05	05/04/2000-06/17/2000	45	MMS_DGOMB
R/V Gyre	00G08	07/29/2000-08/07/2000	47	MMS_NEGOM
Justo Sierra	CANEK_5	06/25/2000-07/03/2000	117	DEEPSTAR
R/V Gyre	01G05	06/02/2001-06/18/2001	13	MMS_DGOMB
Justo Sierra	CANEK_6	05/03/2001-06/14/2001	72	DEEPSTAR
R/V Gyre	02G03	04/02/2002-04/02/2002	1	MMS_DGOMB
R/V Gyre	02G07	06/04/2002-06/13/2002	6	MMS_DGOMB

* MMS_DEEPW: Nowlin et al. (2001); DiMarco et al. (2001)
MMS_LATEXA: Nowlin et al. (1998a, 1998b); Jochens et al. (1998)
MMS_DGOMB: Rowe (2005)
MMS_NEGOM: Jochens and Nowlin (1998, 1999, 2000); Jochens et al. (2002)
DEEPSTAR: CICESE and Mitchell (2002)

Figure 3.1 shows the locations of the stations at which bottle oxygen data have been assembled. These are presented in six parts covering time periods of the order of a decade. Figure 3.2 shows the locations of the stations at which oxygen sensor data have been assembled.

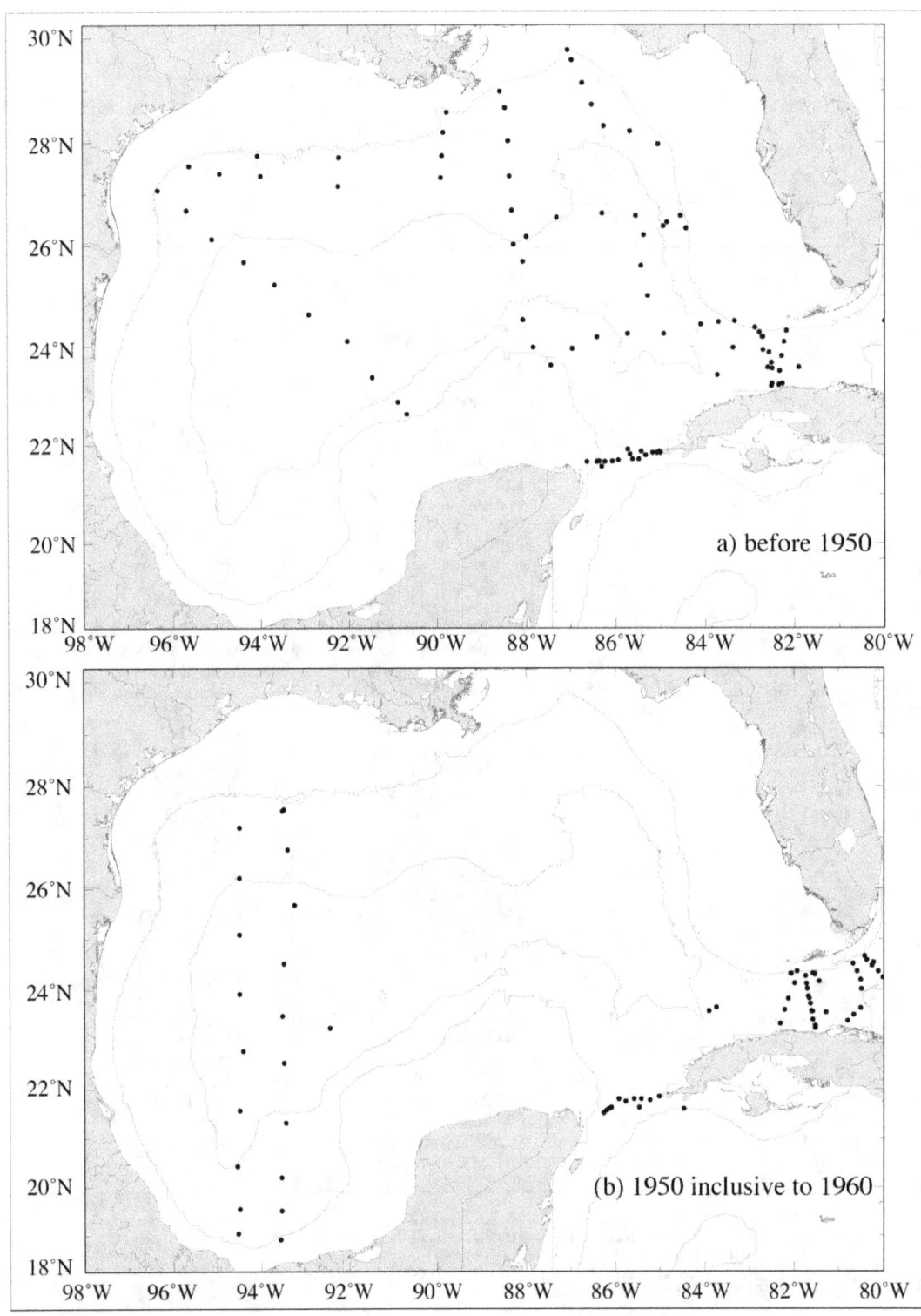

Figure 3.1. Locations of bottle oxygen stations for the deepwater Gulf of Mexico. Shown are stations made (a) before 1950, (b) 1950 to 1960, (c) 1960 to 1970, (d) 1970 to 1980, (e) 1980 to 1990, and (f) 1990 through 2001. Bathymetry contours are 200 m and 3000 m. Not all stations in final data base are shown.

16

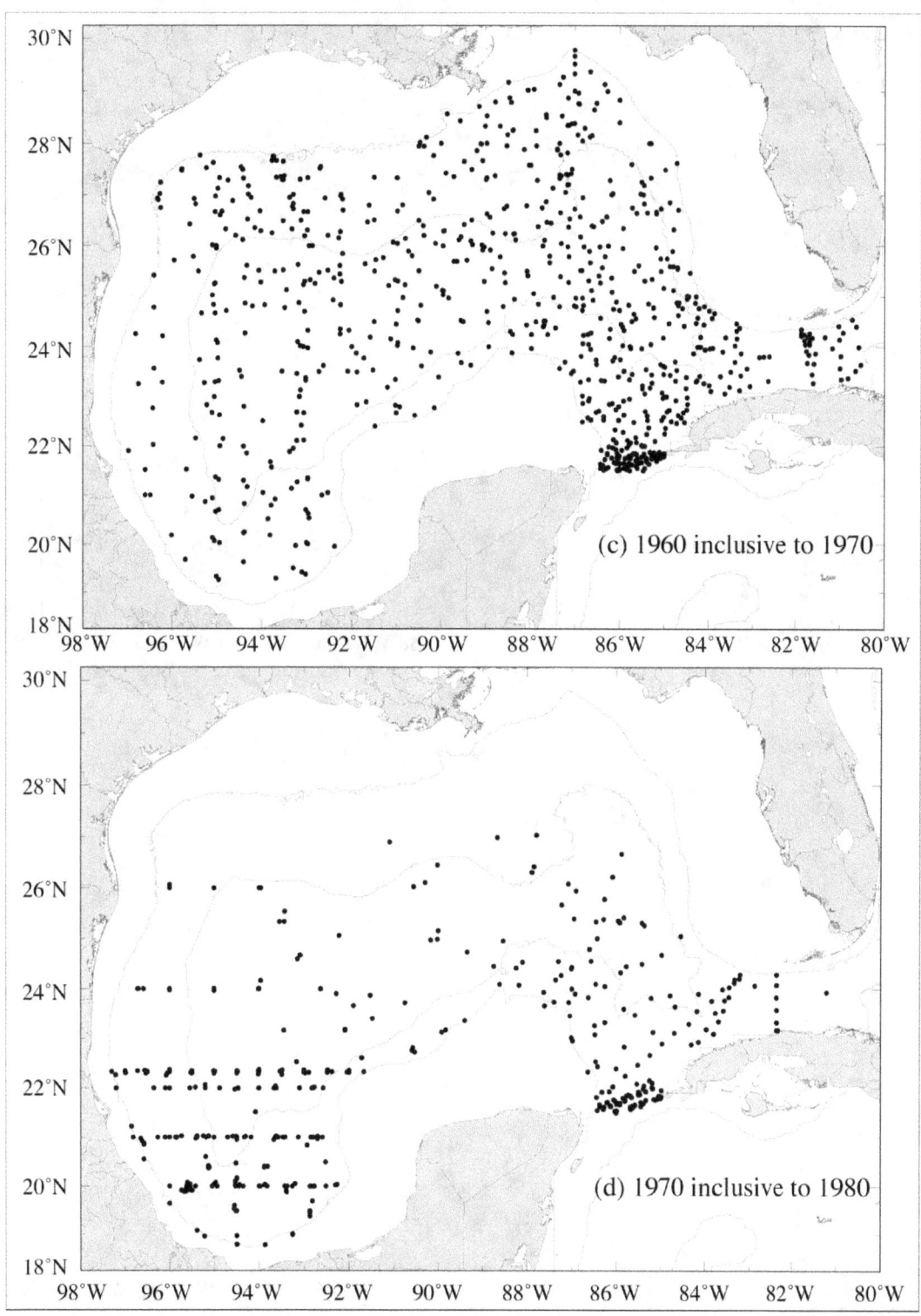

Figure 3.1.   Locations of bottle oxygen stations for the deepwater Gulf of Mexico. (continued)

17

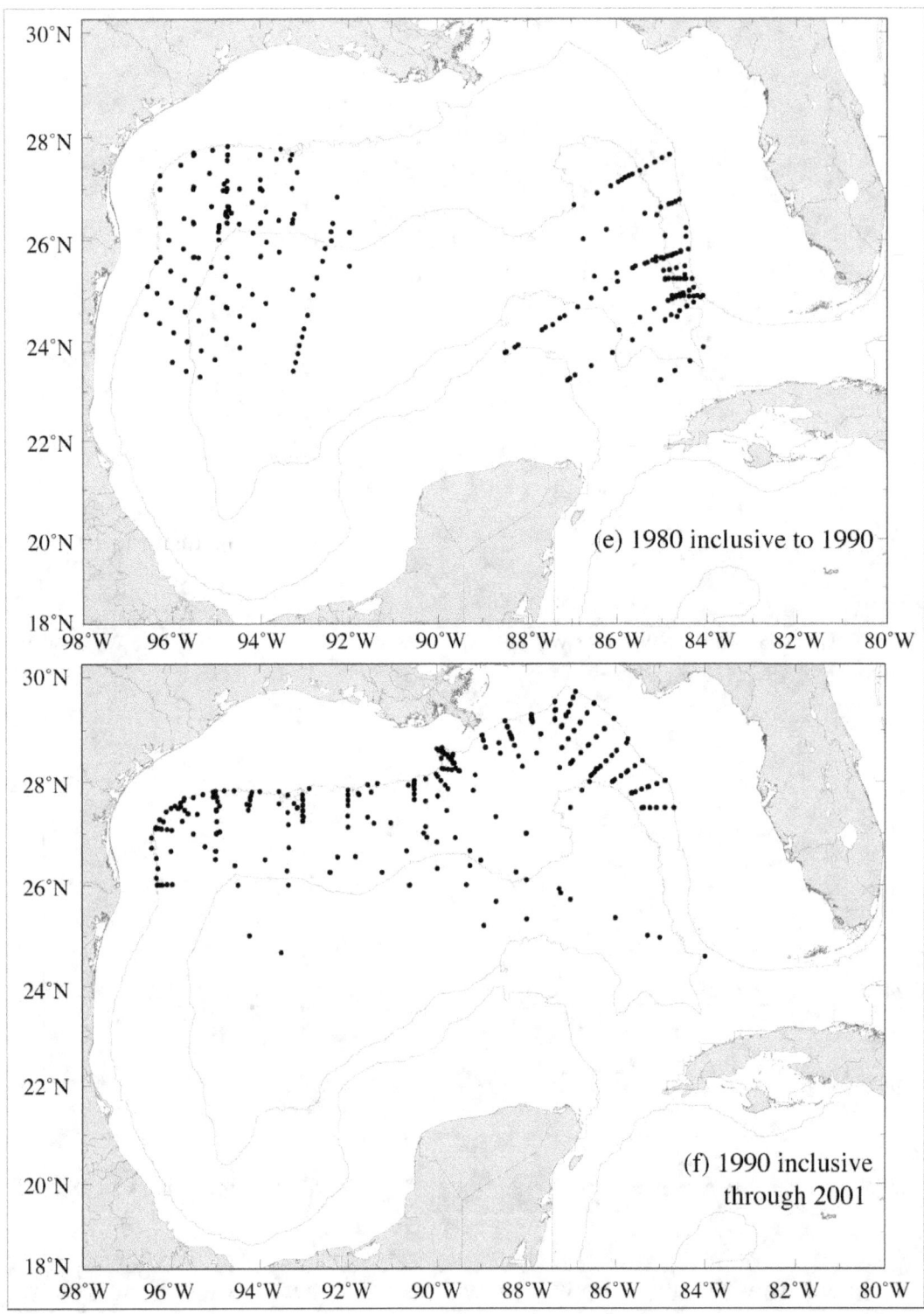

Figure 3.1. Locations of bottle oxygen stations for the deepwater Gulf of Mexico. (continued)

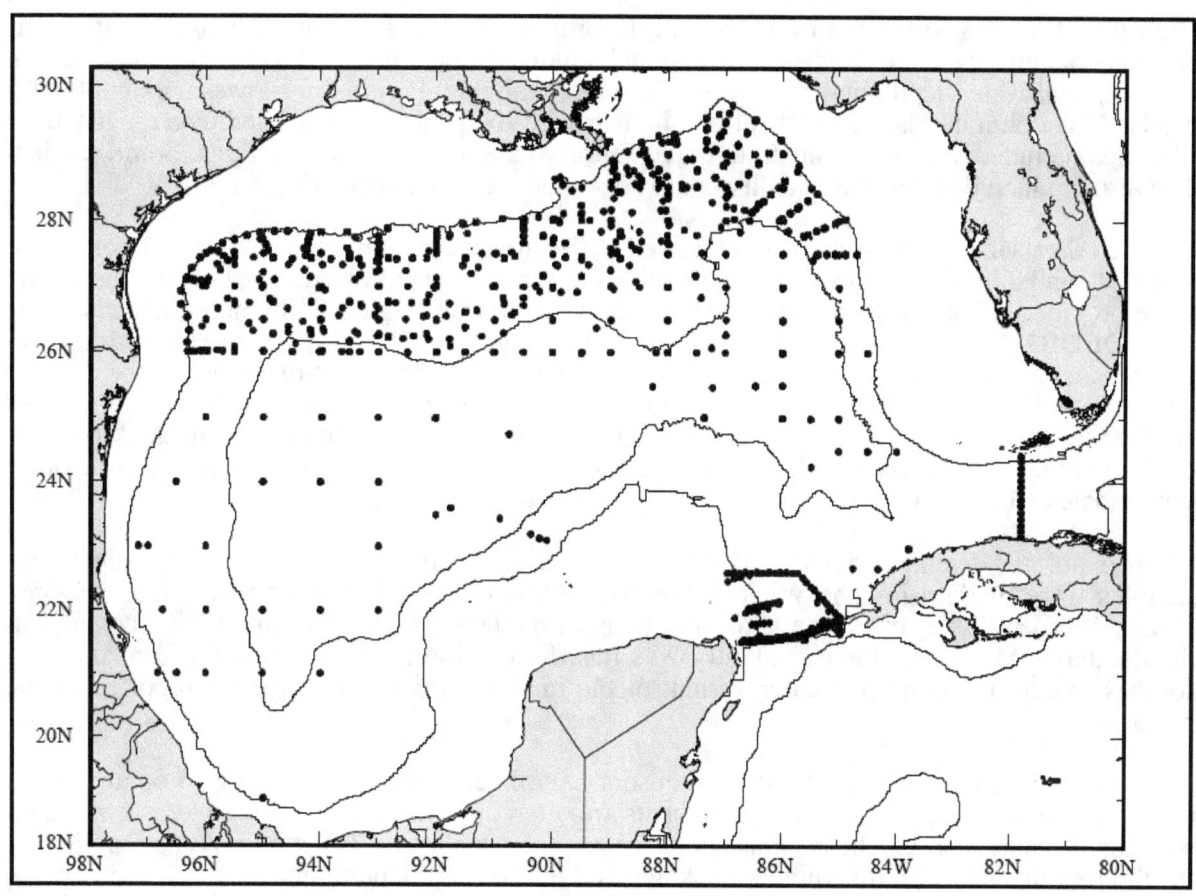

Figure 3.2. Locations of stations with dissolved oxygen sensor data in the deepwater Gulf of Mexico. Bathymetry contours are 200 and 3000 m.

3.2   Data Quality Assurance and Quality Control

Dissolved oxygen data basically fall into two types: 1) discrete values from iodometric analysis (classical Winkler or variants) of water samples collected from Niskin or Nansen bottles, or 2) continuous values from Beckman polarographic oxygen sensors or other sensors mounted on CTD/STD packages. Both data types were assembled. The quality of dissolved oxygen measurements made using the classical Winkler (1888) technique and its variants (e.g., Carpenter 1965a, 1965b, 1966; see also Strickland and Parsons 1972) depends to a large degree on the skill of the analyst and application of analysis techniques. For example, variations in water sampling techniques (including those caused by day/night lighting on the deck and weather conditions under which sampling is done), reagent preparation and use, and visual starch endpoint determinations by different eyes can result in large or small variations in results (see, e.g., Culberson 1991; Culberson et al. 1991) The quality of the polarographic oxygen sensor data, however, depends on the quality of a careful calibration using bottle data. Without such calibration, they are of unknown quality. Most polarographic oxygen sensor data did not show evidence of having been calibrated post-cruise with the appropriate bottle data. Such calibration is a very time consuming process that is beyond the scope of the contract. Therefore, such calibrations were not done on the continuous sensor data. These data were included in the database with available metadata. However, they were not used in the synthesis.

19

Each data file was formatted by station in accordance with a set format, including standard headers containing the metadata. The metadata includes the date/time, latitude, longitude, station number, total water depth, and other information as available. Figure 3.3 shows an example of a header. The columnar data consists of depth, dissolved oxygen, and associated data quality flag. The files include, if known, temperature and salinity data and their quality flags. Some nutrient data also are included, but their quality is unknown and was not checked.

For each data set, the metadata were reviewed for accuracy and completeness. Corrections were made as available information from publications and/or reports allowed. Seafloor depths, either those provided by the data originator or from the U.S. Navy's Digital Bathymetric Data Base 2-min (DBDB2), were compared to the deepest data point in each cast for the given location. Discrepancies were resolved by accepting the bottom depth that was at or exceeded the deepest data point. Where available and not in conflict with the deepest datum, the originator's water depth is reported in the metadata. The longitude and latitude were plotted and checked to see if the locations were reasonable. Dates were examined for reasonableness as compared to other station dates on a given cruise.

Most, if not all, determinations of dissolved oxygen concentrations derived from water-bottle samples were made using the Winkler (1888) method or a variant due to Carpenter (1965a, 1965b, 1966). Winkler-based methods are the usual method used for the time period covered in the database. Moreover, there generally was insufficient information available from the data sources to determine the particular variant of the method applied, so none are reported in the metadata.

Although continuous dissolved oxygen sensors have been developed and are in common use today, data from polarographic oxygen sensors are often questionable due to problems associated with the probe response. The response is nonlinear and sensitive to (a) strong oxygen gradients, (b) the decent rate of the instrument package, and (c) oxygen depletion around the probe tip in non-pumped systems. The sensor also suffers from significant pressure hysteresis (large differences between the upcast and downcast dissolved oxygen profiles). Relatively complex correction algorithms have been developed to correct for these effects (Owens and Millard 1985), but often with unsatisfactory results. The instrument performs best in the deeper portions of casts made in deep water. Any polarographic oxygen sensor data that have not been calibrated using dissolved oxygen bottle data from Winkler-type analytical techniques are unreliable for use in the study analysis. Information was generally lacking on whether such calibrations had been applied. So, the polarographic oxygen sensor data were retained in the Oxygen Archive, but no QA/QC was applied. The data were not used in the reanalysis or synthesis interpretations. Dissolved oxygen sensors that have become available in the last several years may make more reliable measurements (see e.g., the DeepStar data set across the Yucatan Channel); however, the few data sets from these sensors were retained as is in the database with no further evaluation.

All data are retained in the database, including suspect data. The suspect data are handled through the use of flags. A value of zero denotes the datum passes the QA/QC tests. A value of one denotes the datum is suspect, and a two denotes no QA/QC was applied.

Temperature, salinity, and dissolved oxygen data from the Deepwater Archive had undergone QA/QC processing as part of the Deepwater Study. These data had been screened for range and historical envelopes (see Nowlin et al. 2001 for discussion of QA/QC tests applied to temperature, salinity, and dissolved oxygen). The quality flags assigned to these data were retained in the Oxygen Archive, unless further QA/QC applied under this study classified a data point as suspect.

```
Last Modification Date is 04-AUGUST-2003.
The flag, -99.0000, in a data cell means no data are available.
The flag, 0, in a quality cell means data are of good quality.
The flag, 1, in a quality cell means data are suspect.
The flag, 2, in a quality cell means data have not been quality
controlled.
COUNTRY                      : DENMARK
OCL INTERNAL CRUISE CODE     : 8
LATITUDE                     : 21.5700
LONGITUDE                    : -86.3200
DATE (YYYY/MM/DD)            : 1922/02/01
TIME (GMT)                   : 04:00:00
OCL INTERNAL PROFILE CODE    : 20124
NUMBER OF DEPTH LEVELS       : 14
NODC ACCESSION NUMBER        : 7101220
NODC PROJECT CODE            : UNKNOWN
NODC INSTITUTION             : COLLECTION CARLSBERG FOUND. DANA
EXPEDIT.(CHARLOTTENLUND
WOD98 UNIQUE STATION NUMBER  : UNKNOWN
SHIP NAME                    : DANA II
ORIGINATORS CRUISE CODE      : UNKNOWN
ORIGINATORS STATION CODE     : UNKNOWN
ORIGINATOR STATION NUMBER    : 122
SUBMITTING INSTITUTION       : COLLECTION CARLSBERG FOUND. DANA
EXPEDIT.(CHARLOTTENLUND
SUBMITTING PI                : UNKNOWN
DEPTH PRECISION              : UNKNOWN
BOTTOM DEPTH (METERS)        :  1035 as_given
T-S PROBE                    : UNKNOWN
PRIMARY INVESTIGATOR         : UNKNOWN
COLUMN  1 DEPTH              : METERS
COLUMN  2 TEMPERATURE        : DEGREES CELSIUS
COLUMN  3 SALINITY           : PSS or PPT
COLUMN  4 OXYGEN             : MILLILITERS PER LITER
COLUMN  5 NITRITE            : NOT-MEASURED
COLUMN  6 NITRATE            : NOT-MEASURED
COLUMN  7 PHOSPHATE          : NOT-MEASURED
COLUMN  8 SILICATE           : NOT-MEASURED
COLUMN  9 SALINITY II        : NOT-MEASURED
COLUMN 10 T-S QUALITY FLAG   : SUSPECT = 1 WHEN SIGMA-THETA > 2.3 std.dev.
FROM HISTORICAL MEAN, 0 = PASSED, 2 = UNTESTED
COLUMN 11 OXY QUALITY FLAG   : SUSPECT = 1 WHEN OXYGEN      > 2.3 std.dev.
FROM HISTORICAL MEAN, 0 = PASSED, 2 = UNTESTED
COLUMN 12 NO2 QUALITY FLAG   : SUSPECT = 1 WHEN NITRITE     > 2.3 std.dev.
FROM HISTORICAL MEAN, 0 = PASSED, 2 = UNTESTED
COLUMN 13 NO3 QUALITY FLAG   : SUSPECT = 1 WHEN NITRATE     > 2.3 std.dev.
FROM HISTORICAL MEAN, 0 = PASSED, 2 = UNTESTED
COLUMN 14 PO4 QUALITY FLAG   : SUSPECT = 1 WHEN PHOSPHATE   > 2.3 std.dev.
FROM HISTORICAL MEAN, 0 = PASSED, 2 = UNTESTED
COLUMN 15 SIO3 QUALITY FLAG  : SUSPECT = 1 WHEN SILICATE    > 2.3 std.dev.
FROM HISTORICAL MEAN, 0 = PASSED, 2 = UNTESTED
*END*
```

Figure 3.3.   Example of a data file header for the dissolved oxygen bottle data.

The QA/QC of the bottle data consisted of three tiers of checks. The primary QA/QC focused on removal of duplicate data sets and the flagging of unrealistic wild point outliers. The secondary QA/QC focused on examination of the bottle data for and flagging of unreasonable data points as defined by statistical limits based on standard deviations. The tertiary QA/QC consisted of identification of problem points during the analysis and synthesis phase of the study. Each tier is discussed briefly below.

Primary QA/QC: Checks were made for duplicate data sets, with duplicates being removed. Data were plotted by depth, sigma-theta level, or potential temperature and examined for overall reasonableness. Physically unrealistic data points (spikes or wild points) or profiles were flagged as suspect.

Secondary QA/QC: This tier of QA/QC used the database to develop a mean and standard deviation for binned data. Data that were outside 2.3 standard deviations (2.3σ) from the mean based on depth were flagged as suspect. A histogram of data by depth was constructed (Figure 3.4) to identify depth bins for determination of means and standard deviations. The depths bins selected were: 50-m increments from 0 to 1000 m depths; 250-m increments from 1000 to 2000 m; 500-m increments from 2000 to 3000 m; and a bottom bin including all data below 3000 m. Table 3.3 gives the statistics that were used. These were determined for the whole Gulf based on the uncorrected data available in January 2003 from over 2200 stations. Figure 3.5 shows the mean and 2.3σ used for this step.

Tertiary QA/QC: The tertiary QA/QC step was a product of the analysis and synthesis phase of the study. The few data points that were determined to be unreliable during this phase of the study were flagged as suspect.

Discussion of QA/QC results: Presented here are several examples of data sets and how they fit within the wild point, 2.3σ, and suspect profile criteria. These examples are intended to alert the users of the database to beware of and use the quality flags for the data points and to examine the data themselves to determine whether additional QA/QC should be applied for the intended use.

First is shown one of the data sets identified by Nowlin et al. (1969) as being of good quality. These data are from a summer 1967 *Alaminos* cruise in the eastern and central Gulf. Station locations are shown in the top panel of Figure 3.6. The bottom panels of Figure 3.6 show the data by depth and grouped by east and west locations, divided at –90°W longitude. Note the characteristic low level of scatter in the data below 1500 m that was noted by Nowlin et al. (1969). Compare the larger scatter of the winter 1962 *Hidalgo* data located below 1500 m, that were noted by Nowlin et al. (1969) to be of questionable quality (Figure 3.7). The data that lie outside the 2.3σ contours are readily apparent in these plots. However, Nowlin et al. (1969) state that the large scatter seen at the various depths, which includes that scatter within 2.3σ contours shown, makes all dissolved oxygen data from this cruise suspect. The Oxygen Archive, however, does not have these data flagged as suspect unless they are outside the 2.3σ limits.

A more recent data set was collected over the northern Gulf aboard the R/V *Gyre* in May/June 2000. Station locations are given in the top panel of Figure 3.8. Note, in the bottom panels of the figure that the scatter below 1500 m is similar to that of the 1967 *Alaminos* cruise. The profiles also suggest that the profile, indicated by the arrow, from a station in the western Gulf may be suspect. After further inspection, the cast was retained in the data base but was excluded from the set of profiles used to form the canonical set (see Section 3.3 below).

Many of the data sets do not sample below 1500 m, even when the water depth is much greater. An example is shown in Figure 3.9. The stations are located in the southwestern Gulf, and the

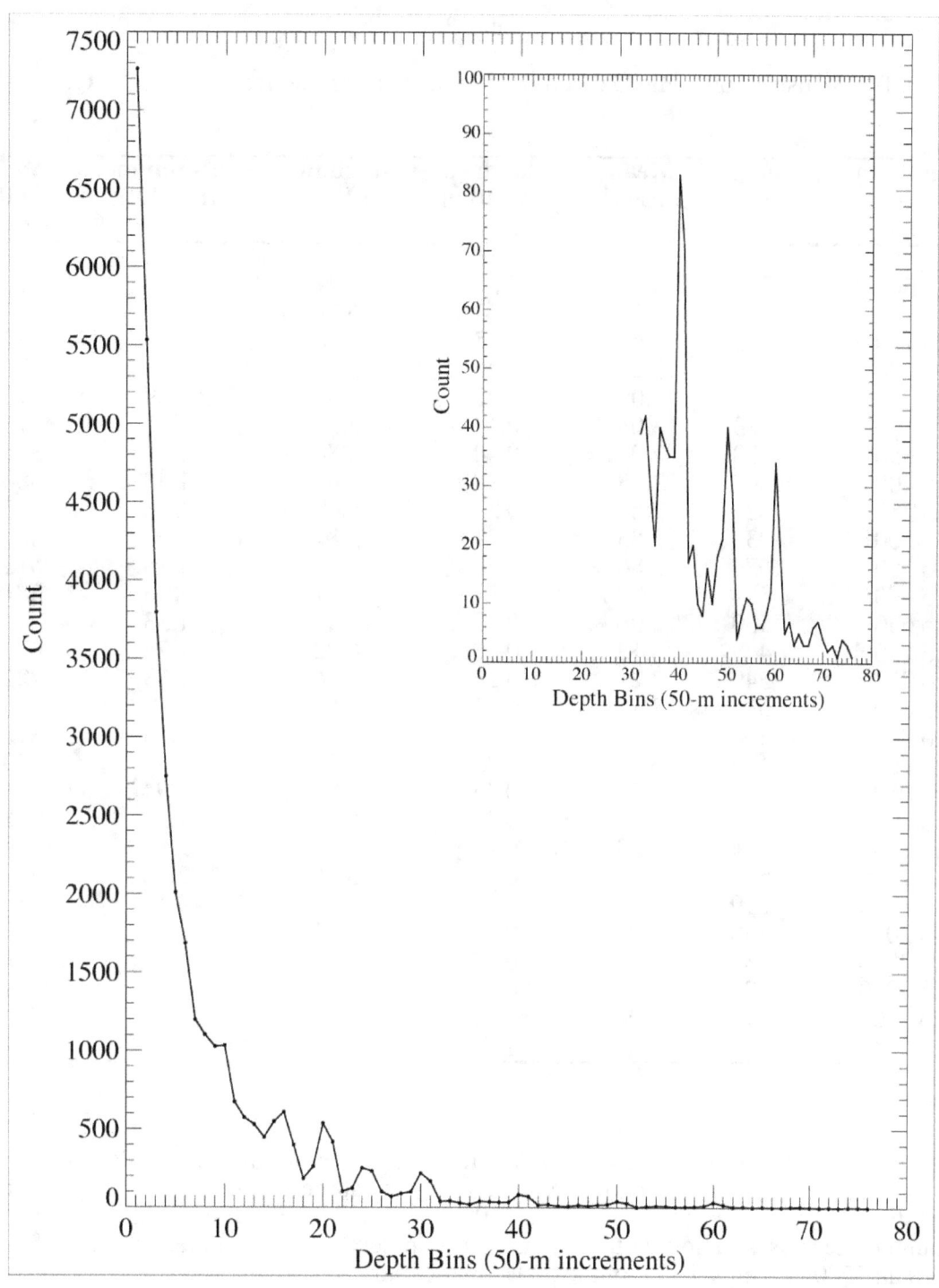

Figure 3.4.   Histogram of bottle oxygen data count in 50-m bin increments. Inset shows a close-up of the counts for bins 32-76.

23

Table 3.3

Bin Statistics for Bottle Oxygen Data in the Gulf of Mexico Used in QA/QC

Bin interval (m)	No. Points	Mean (mL·L⁻¹)	Standard Deviation (mL·L⁻¹)	Maximum (mL·L⁻¹)	Minimum (mL·L⁻¹)	Mean Depth of Data (m)
0 to 50	7263	4.69	0.37	7.47	2.08	18.9
50 to 100	5533	4.37	0.63	6.79	1.63	73.2
100 to 150	3795	3.61	0.65	6.16	1.77	122.4
150 to 200	2750	3.28	0.48	5.49	2.05	173.5
200 to 250	2009	3.09	0.42	4.99	1.89	218.5
250 to 300	1683	3.03	0.47	5.58	1.52	275.9
300 to 350	1198	2.94	0.44	4.86	1.51	314.9
350 to 400	1102	2.88	0.40	4.69	1.48	380.9
400 to 450	1028	2.83	0.33	4.09	1.26	415.5
450 to 500	1033	2.85	0.31	5.81	1.73	482.0
500 to 550	674	2.86	0.30	6.18	2.18	513.5
550 to 600	574	2.92	0.25	4.31	1.96	581.3
600 to 650	530	2.96	0.28	4.46	1.53	614.0
650 to 700	449	3.13	0.39	6.82	1.78	678.4
700 to 750	549	3.21	0.28	4.36	2.13	722.0
750 to 800	609	3.34	0.37	4.65	1.49	775.8
800 to 850	400	3.47	0.38	4.43	1.18	812.0
850 to 900	185	3.64	0.45	6.75	2.10	879.2
900 to 950	263	3.77	0.35	4.81	1.94	921.7
950 to 1000	536	3.99	0.35	5.53	1.89	982.2
1000 to 1250	1134	4.33	0.42	7.16	2.56	1109.3
1250 to 1500	581	4.73	0.33	5.56	2.66	1398.6
1500 to 1750	299	4.86	0.31	5.63	3.15	1562.8
1750 to 2000	230	4.94	0.42	5.95	1.88	1893.3
2000 to 2500	231	5.04	0.37	6.02	3.19	2218.0
2500 to 3000	128	5.00	0.30	5.66	3.37	2761.5
3000 to 4000	76	4.94	0.31	5.54	3.84	3274.3

data were collected in fall 1970 from the Mexican ship *Virgilio Uribe*. This lack of sampling below 1500 m was a common occurrence, particularly in data sets from 1970 to the present. Only 10% of stations in water depths greater than 1500 m sampled 95% of the available water column. This number decreases rapidly with increasing ocean depth. For example, only 5% of stations deeper than 2500 m sampled 95% of the total water depth.

There also are a number of data sets that are unusual, but which were not flagged as suspect. One "discrete" data set that was examined was from the 1970 *Tursiops* cruise near the region of the Loop Current inflow (Figure 3.10). Each profile consists of far more data than is generally collected by discrete samplers, and we know of no probe-derived oxygens from as early as 1970. It is likely this is a case of data that were interpolated to standard depths, rather than actual data. But there is insufficient information available to determine this for certain.

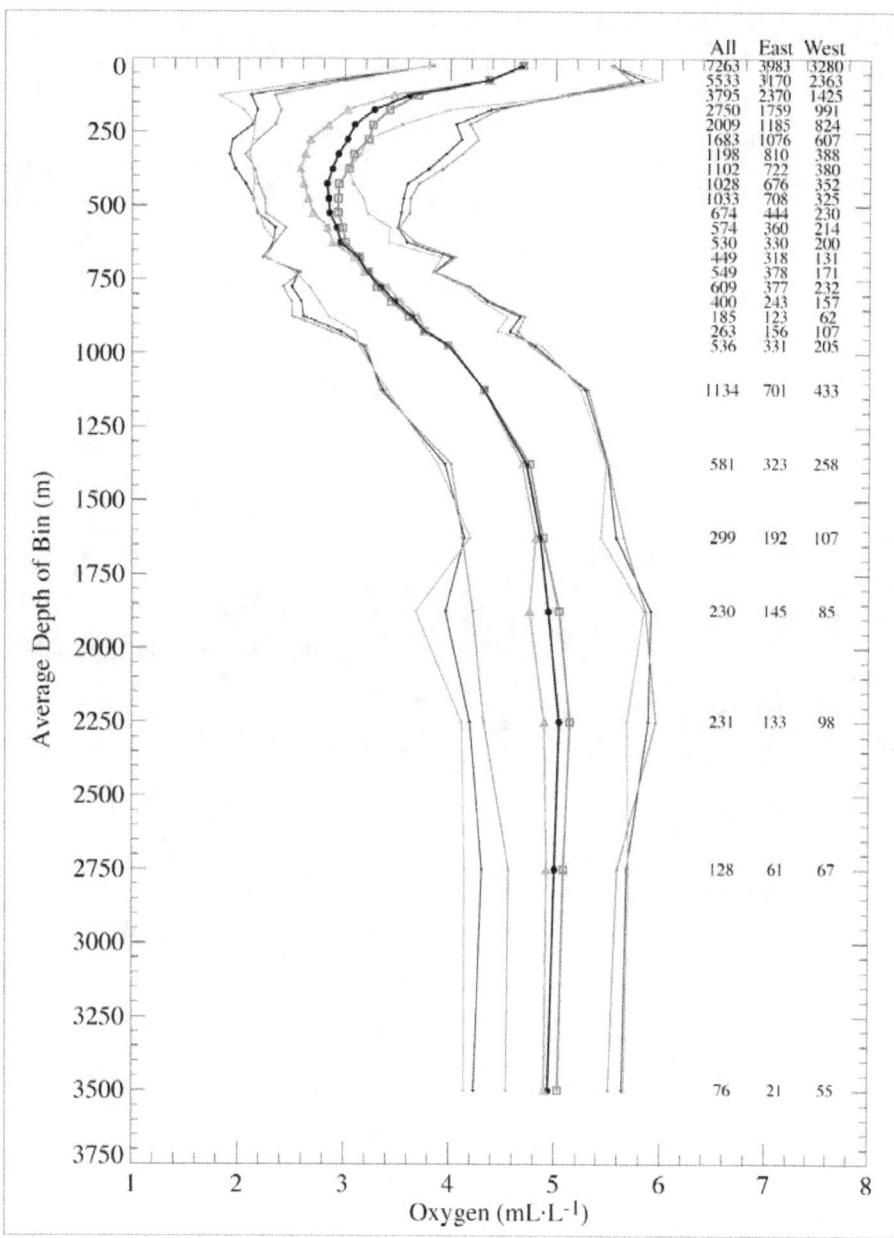

	All	East	West
	7263	3983	3280
	5533	3170	2363
	3795	2370	1425
	2750	1759	991
	2009	1185	824
	1683	1076	607
	1198	810	388
	1102	722	380
	1028	676	352
	1033	708	325
	674	444	230
	574	360	214
	530	330	200
	449	318	131
	549	378	171
	609	377	232
	400	243	157
	185	123	62
	263	156	107
	536	331	205
	1134	701	433
	581	323	258
	299	192	107
	230	145	85
	231	133	98
	128	61	67
	76	21	55

Figure 3.5. Oxygen statistics by depth bin. Shown are the mean and 2.3σ statistics for the whole Gulf (black; circles), west of 90°W (red; squares), and east of and including 90°W (green; triangles). Symbols show the middle of each depth bin. Numbers on the right give the number of data points for each bin. The mean depths of the oxygen minimum layer are approximately 415 m for the whole Gulf, 525 m for the eastern Gulf, and 400 m for the western Gulf.

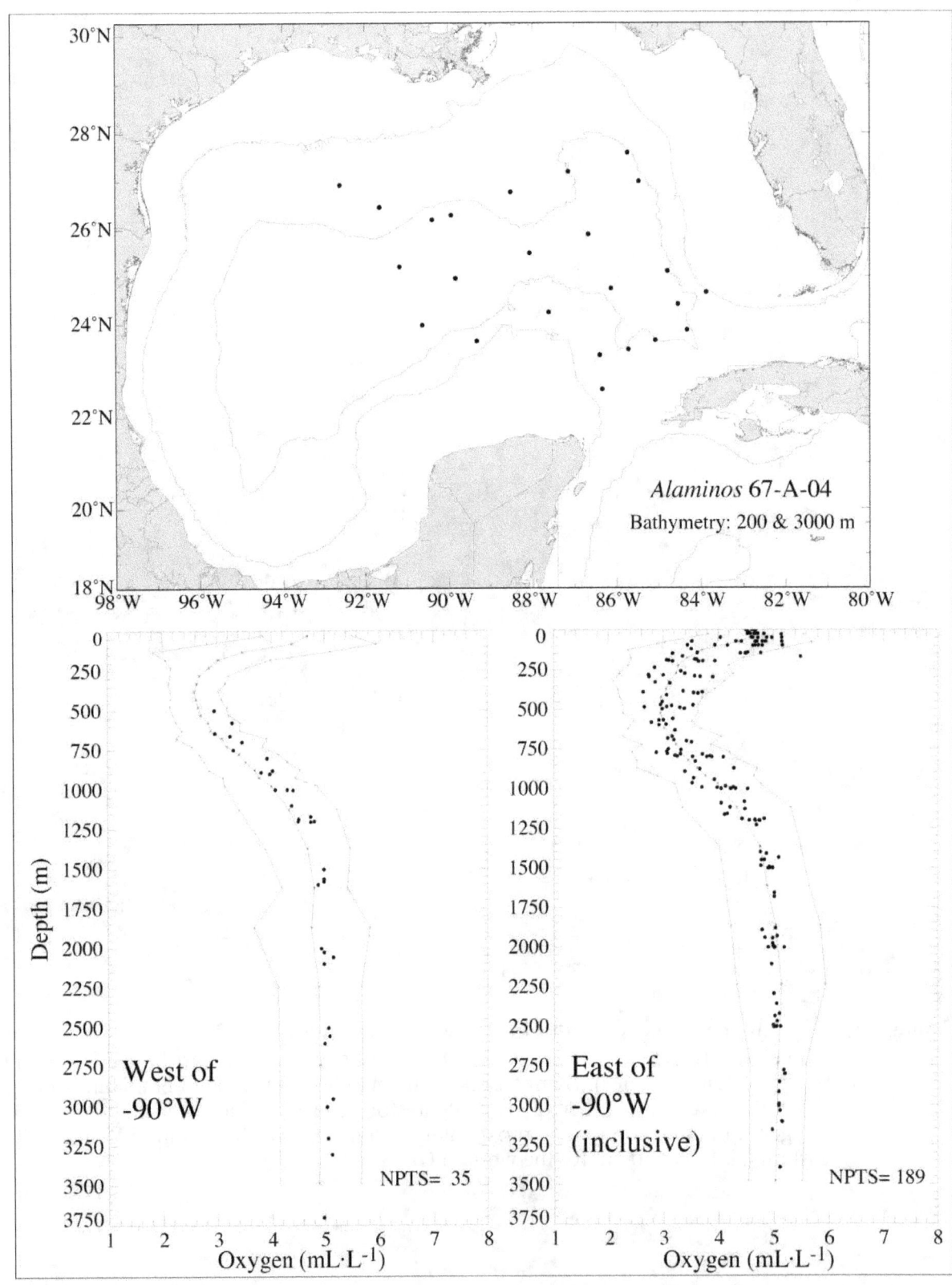

Figure 3.6. Locations of stations (top) and oxygen data (bottom) for R/V *Alaminos* cruise, 67-A-04, on 11 June - 1 July 1967. Historical mean and 2.3σ are shown.

26

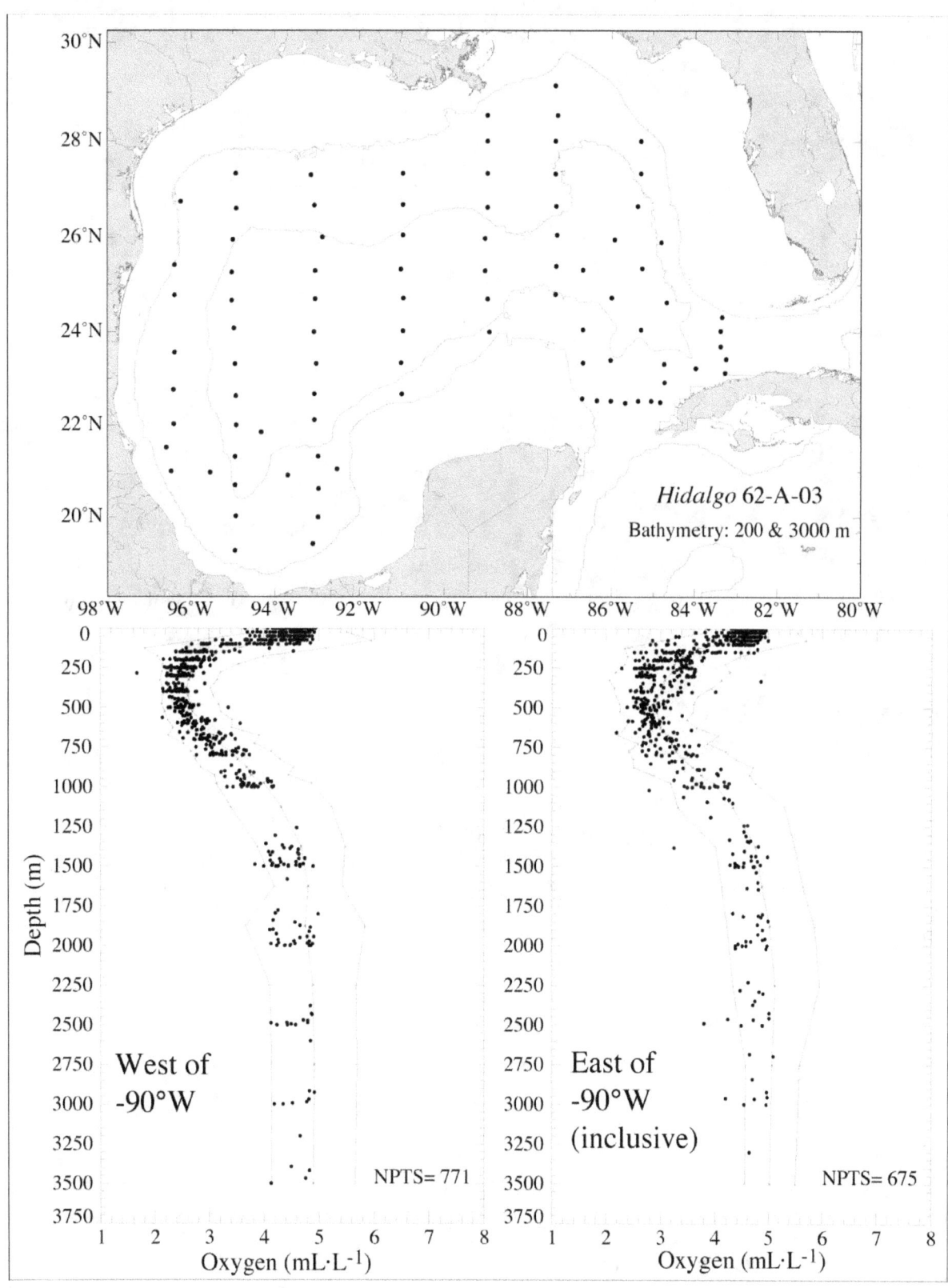

Figure 3.7. Locations of stations (top) and oxygen data (bottom) for R/V *Hidalgo* cruise, 62-H-03, on 14 February - 31 March 1962. Historical mean and 2.3σ are shown.

27

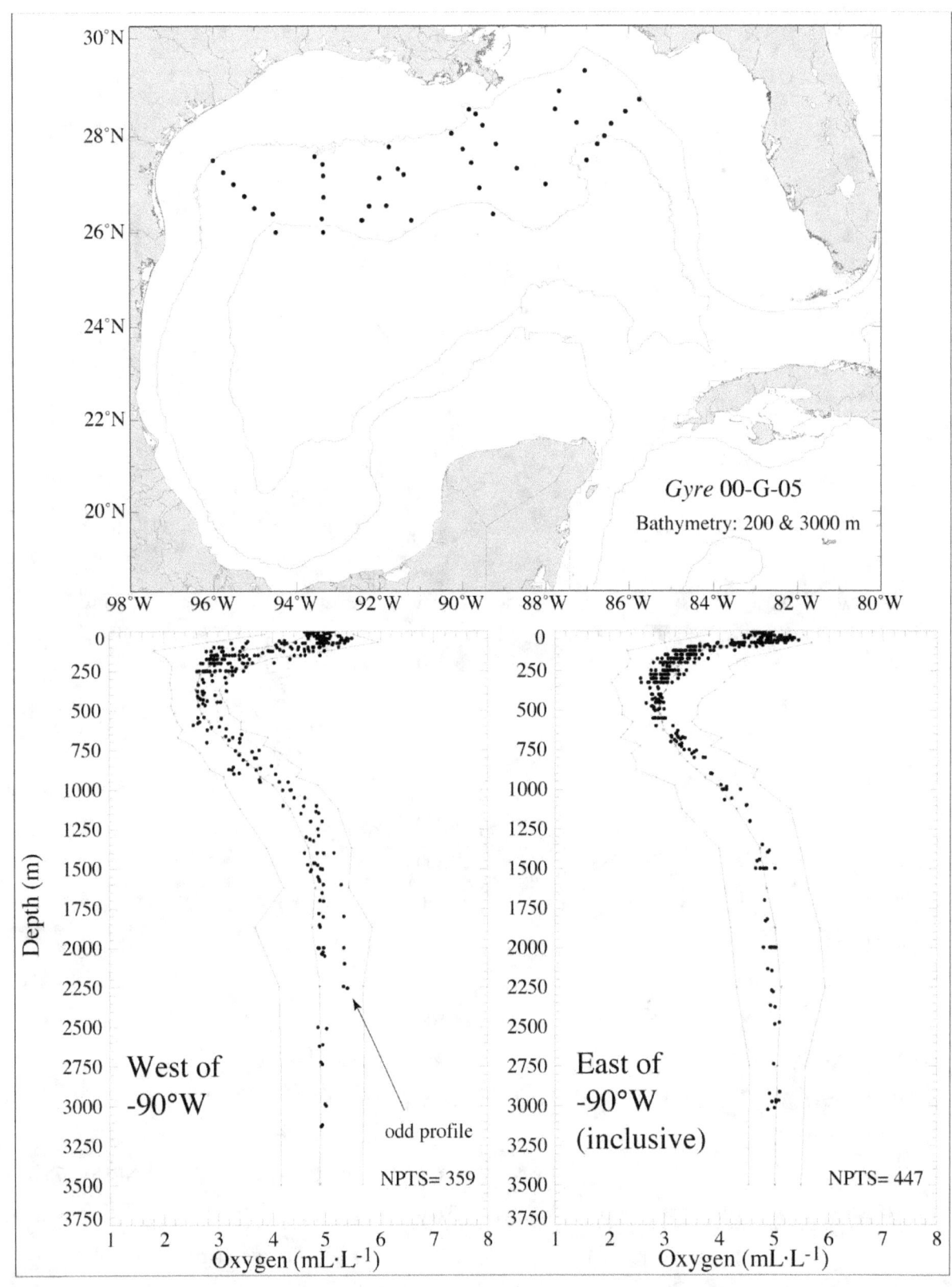

Figure 3.8. Locations of stations (top) and oxygen data (bottom) for R/V *Gyre* cruise, 00-G-05, on 4 May - 17 June 2000. Historical mean and 2.3σ are shown.

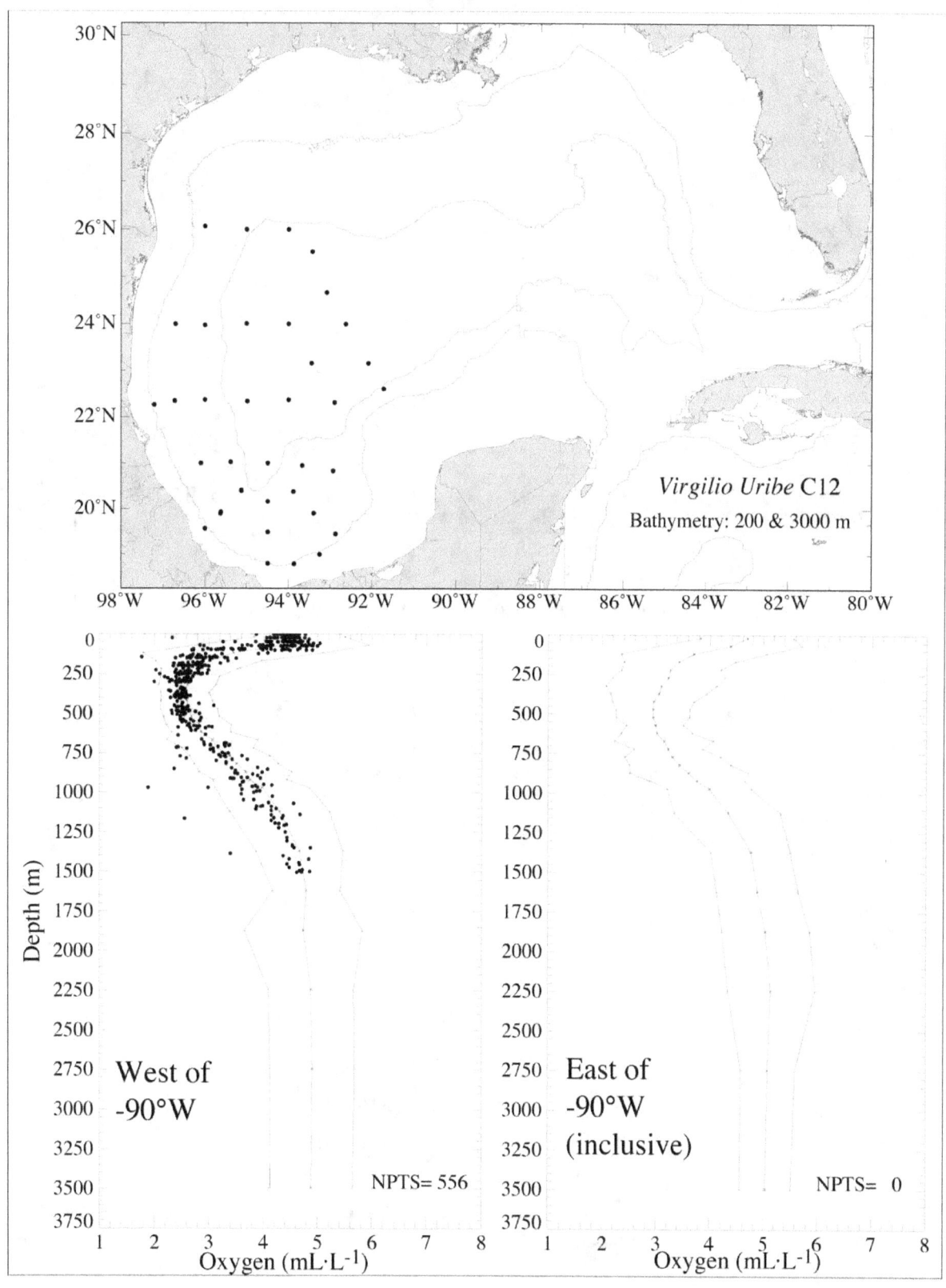

Figure 3.9. Locations of stations (top) and oxygen data (bottom) for the *Virgilio Uribe* cruise, C12, on 31 October - 13 November 1970. Historical mean and 2.3σ are shown.

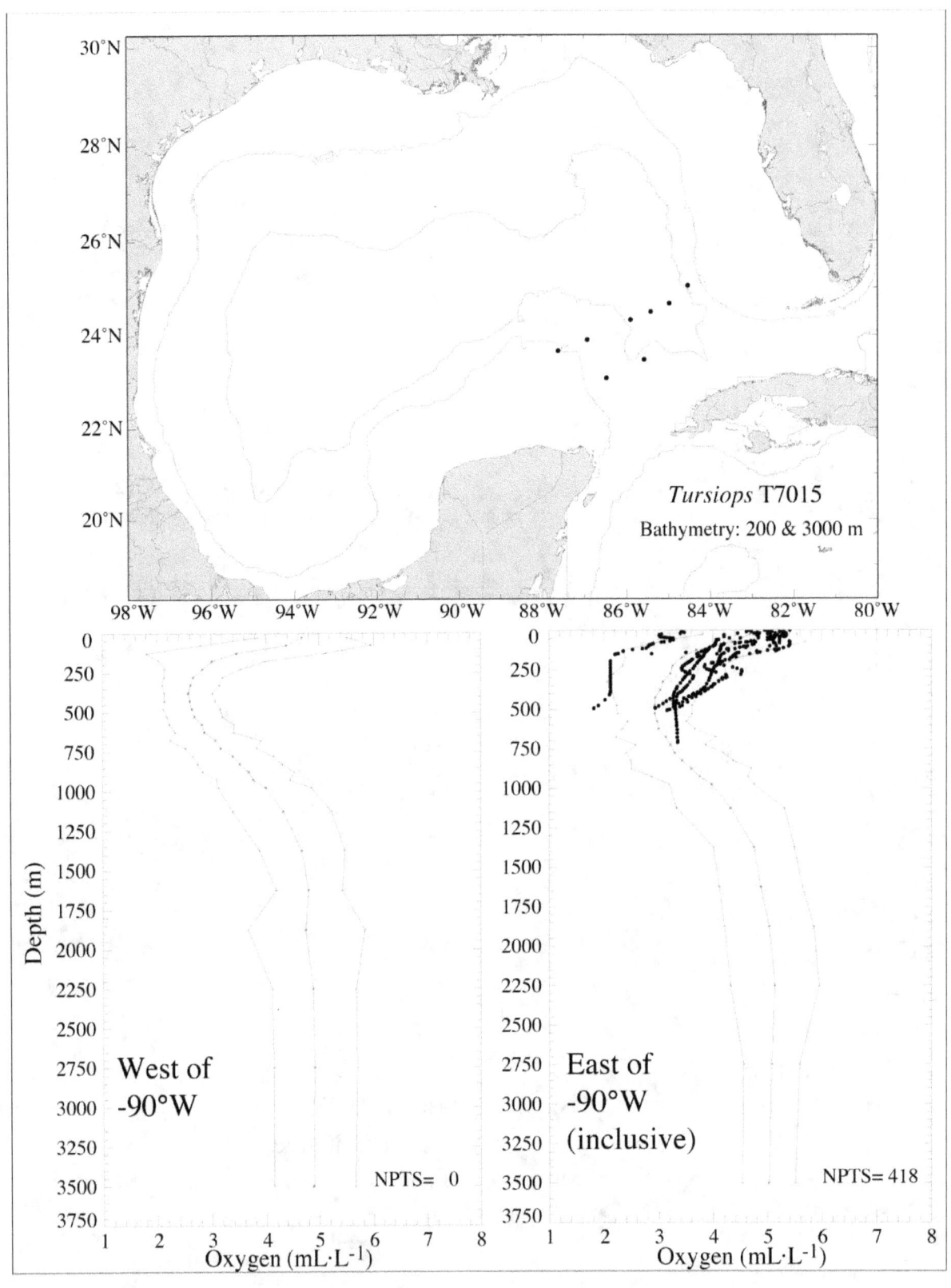

Figure 3.10. Locations of stations (top) and oxygen data (bottom) for the *Tursiops* cruise, T7015, on 2 - 11 May 1970. Historical mean and 2.3σ are shown.

The temporal nature of the sampling was examined to determine if there was bias in the seasons in which samples were taken. Table 3.4 shows the number of stations with dissolved oxygen samples from bottles by decade and month. Shown are both the numbers for all the stations in the whole Gulf and for stations that had at least one sample taken at or below 1500 m depth. Sampling in the whole Gulf and at or deeper than 1500 m was conducted, respectively, 34 and 40% of the time in spring (March-May), 25 and 20% in summer (June-Aug), 26 and 21% in fall (Sept-Nov), and 15 and 16% in winter (Dec-Feb). Although summer is most frequently sampled and winter least, the distribution in time is reasonable for analysis purposes.

Table 3.4

Number of Stations With Bottle Dissolved Oxygen by Month and Decade

Decade	Jan	Feb	Mar	Apr	May	Jun	Jul	Aug	Sep	Oct	Nov	Dec	Total
All stations													
<1950	0	29	33	24	5	0	0	0	0	0	0	0	91
1950s	28	42	11	23	22	8	41	28	18	10	0	27	258
1960s	38	153	92	70	72	146	56	155	176	21	27	4	1010
1970s	65	11	16	130	262	49	24	30	13	126	172	0	898
1980s	66	64	44	74	64	12	40	18	11	55	80	13	541
1990s	14	15	8	58	189	29	62	132	8	43	219	0	777
2000+	0	0	0	48	26	31	14	53	9	1	0	11	193
Total	211	314	204	427	640	275	237	416	235	256	498	55	3768
Stations with samples collected at or below 1500 m													
<1950	0	8	6	3	3	0	0	0	0	0	0	0	20
1950s	16	23	6	8	5	3	2	14	9	0	0	2	88
1960s	14	19	41	28	15	52	18	13	22	3	1	0	226
1970s	0	0	4	64	58	3	0	1	1	40	37	0	208
1980s	29	3	4	0	1	0	0	0	0	1	16	2	56
1990s	2	0	0	0	0	1	1	1	0	1	0	0	6
2000+	0	0	0	0	11	16	0	0	0	0	0	0	27
Total	61	53	61	103	93	75	21	29	32	45	54	4	631

3.3   Box Model Initialization Data Sets

Two sets of canonical profiles of dissolved oxygen, temperature, and salinity were developed for use in initialization of the box model discussed in Section 5. One was for the Yucatan Channel; the other was for the deep Gulf basin. All oxygen values were based on Winkler titrations, i.e., no continuous probe values were used. The Yucatan Channel profile was constructed from profiles collected within the box with corners at 21°N, 87°W and 22°N, 84.3°W. Table 3.5 shows values for the Yucatan Channel; Table 3.6 shows values for the deep Gulf basin.

Table 3.5

Yucatan Channel Canonical Profiles for Temperature, Salinity, and Dissolved Oxygen

Depth (m)	Potential Temperature (°C)	Salinity	Dissolved Oxygen (mL·L^{-1})
0.	27.47	35.95	4.56
10.	27.39	36.06	4.56
100.	24.64	36.32	4.15
200.	19.44	36.53	3.57
500.	11.31	35.31	3.00
800.	6.41	34.41	3.33
1150.	4.49	34.49	4.50
1500.	4.15	34.97	4.87
1750.	4.06	34.97	5.09
2000.	3.98	34.98	5.43

Table 3.6

Deep Gulf Basin Canonical Profiles for Temperature, Salinity, and Dissolved Oxygen

Depth (m)	Potential Temperature (°C)	Salinity	Dissolved Oxygen (mL·L^{-1})
0.	26.83	35.71	4.65
10.	26.38	35.72	4.74
100.	20.29	36.40	3.68
200.	15.41	36.00	3.06
500.	8.67	35.06	2.84
800.	5.68	34.90	3.61
1150.	4.45	34.95	4.48
1500.	4.12	34.97	4.91
1750.	4.07	34.97	4.95
2000.	4.04	34.97	4.99
3000.	3.99	34.97	5.08
3500.	3.97	34.97	4.90

The canonical dissolved oxygen profiles were constructed through a series of steps designed to eliminate bad cruises, bad profiles, and bad points from the data base leaving only the best-of-the-best data. These data then were averaged by depth, smoothed, spline-fit to 1-m intervals, and smoothed again to produce the final representative profiles. Each profile was considered three times. Only profiles that were accepted all three times were used to compute the canonical profiles.

In each pass, the ensembles of dissolved oxygen versus depth and dissolved oxygen versus potential temperature were plotted in adjacent panels. Individual profiles were over-plotted in a different color one at a time and were accepted or rejected by eye. After the initial screening, the results were examined by cruise. Cruises with large numbers of rejected profiles were identified and all profiles from that cruise were removed from further consideration. In the second and third screening, individual profiles were eliminated on a profile-by-profile basis by comparing them to the ensemble plot of profiles that had passed the preceding step. Only profiles that were accepted three times were used in the averaging and smoothing steps. The number of profiles that passed the three screens was 161 for the Yucatan Channel and 906 for the deep Gulf basin.

The QA/QC processing of the temperature, salinity, and dissolved oxygen data resulted in the assignment of a quality control flag to each data point. If any of the temperature, salinity or dissolved oxygen values at a given depth within an accepted profile were flagged bad, then all three of the values were rejected for use in the canonical data set. The remaining data then were averaged by depth and smoothed with sequential applications of a moving box car of 3, 5, 7, 11, and 33 points. The resulting profile was spline-fit to 1-m bins and smoothed one final time with a 33-point moving average.

4 DISSOLVED OXYGEN IN THE DEEPWATER GULF OF MEXICO

Dissolved oxygen is one of the most intensely studied components of seawater. It is essential for life to many marine organisms. Broecker (e.g., 1974) has categorized it as being intermediate in limiting oceanic biologic processes. The dissolved oxygen concentration can vary, even on local scales, from undetectable (anoxic) to supersaturated with respect to atmospheric concentration. Biological processes both produce oxygen via photosynthesis and consume it via respiration. Abiotic redox reactions also can lead to oxygen consumption. Physical processes such as air-sea exchange and benthic oxygen demand may cause further changes in oxygen concentration. Different water masses have differing oxygen concentrations depending on their source and history. Movement and mixing of different water masses therefore give further texture to the distribution of oxygen in the ocean. For most major ocean basins this leads to the "classic" dissolved O_2 depth profile shown in Figure 4.1. In this section, these biological, chemical and physical processes, which control the concentration and distribution of dissolved oxygen, are examined in the Gulf of Mexico based on our general understanding of these oceanic processes and the data we have gathered as part of this project.

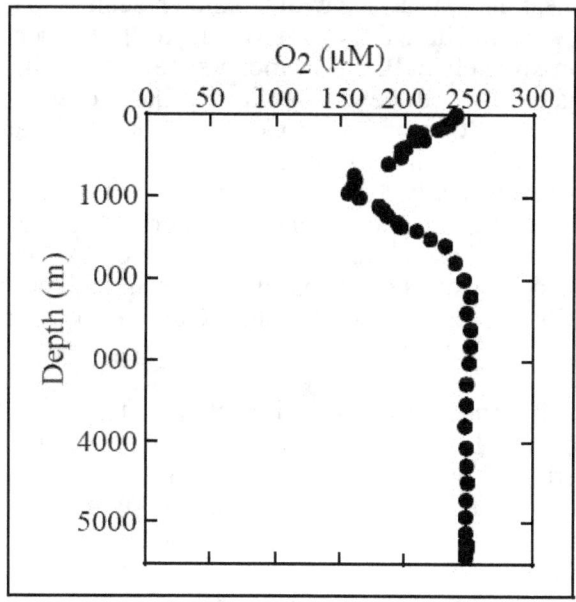

Figure 4.1.   Dissolved O_2 in the North Atlantic Ocean (based on Millero 1996, Fig. 6.9).

Our discussion starts with Section 4.1, which presents background information on the physical oceanography, circulation, and water masses of the Gulf of Mexico and a discussion of flushing time. These processes set the major features of dissolved oxygen distribution in the Gulf of Mexico. The brief review presented here is intended to highlight the physical oceanographic setting; it is not intended to be a comprehensive review (see Nowlin et al. 2000). The sources and sinks of dissolved oxygen in the deepwater of the Gulf of Mexico are identified in Section 4.2, and the operative physical and biogeochemical processes are discussed. Section 4.3 presents the results of a reanalysis of the historical oxygen database in the deepwater Gulf of Mexico. The potential effects of oil and gas discharges are considered in Section 4.4.

4.1   Physical Oceanographic Background

The distribution of dissolved oxygen in the water column of the deepwater Gulf of Mexico (hereinafter "Gulf") is influenced by circulation, which transports waters and the oxygen they contain throughout the Gulf. The mean circulation in the upper 1000 m has been observed to be generally anticyclonic, except perhaps in the Bay of Campeche, with many cyclonic and anticyclonic eddies present at any given time (Figure 4.2), while sparse data sets and model results (Figure 4.3) suggest the mean circulation in the deeper waters is cyclonic, but also with many cyclonic and anticyclonic eddies present (e.g., data: Molinari et al. 1978; Hofman and Worley 1986; Hamilton 1990; Hamilton and Lugo-Fernandez 2001; Nowlin et al. 2001; models: Welsh and Inoue 2000; Sturges et al. 2004). Water masses are not formed in the Gulf of Mexico, but water masses from the Atlantic move into the Gulf through the Yucatan Channel. With a sill depth of approximately 2000 m, this passage is deep enough to allow deep water masses, which have relatively high dissolved oxygen concentrations, to enter the Gulf. These are the source of replenishment of dissolved oxygen to the deep Gulf waters. This section explores these factors in more detail.

4.1.1   Brief Review of Physical Oceanography of the Deepwater Gulf of Mexico

Reviewing the historical current database, Nowlin et al. (2001) found that the highest maximum and mean current speeds were near the sea surface with maxima reaching up to 200 cm·s^{-1} in the eastern Gulf and 100 cm·s^{-1} in the western Gulf. They also found that speeds decreased with depth, tending toward a minimum of approximately 5-10 cm·s^{-1} mean speed at depths of 800-1000 m, and that speeds increased somewhat with depth below that level, likely due to bottom intensification of currents (see also Molinari and Morrison 1988; Cooper et al. 1990; and Hamilton et al. 1997). Maximum currents below 1500 m have been observed to reach 20-50 cm·s^{-1} and for a few measurement locations to reach nearly 100 cm·s^{-1} (Nowlin et al. 2001). Hamilton and Lugo-Fernandez (2001) found recurring episodes of energetic currents of 40-50 cm·s^{-1}, with maximum velocities sometimes exceeding 85 cm·s^{-1}, about 10 m above bottom in ~2000 m depth near the Sigsbee Escarpment. Thus, deep waters of the Gulf are not quiescent, and the oxygen-rich deep waters from the Yucatan Channel and the southeastern Gulf can be carried into the northern and southwestern Gulf to renew those waters.

There are two principal forcing functions of the circulation in the upper 800 to 1000 m: the Loop Current system, including the Loop Current itself, Loop Current eddies (LCEs) and other circulation phenomena derived from them (Figure 4.4), and wind stress (overviews are given in Nowlin et al. 2001 and Continental Shelf Associates, Inc. 2000). The Loop Current system begins with the Yucatan Current that enters the Gulf from the Caribbean Sea through the Yucatan Channel. It then separates from the Campeche Bank, becoming the Loop Current. The Yucatan Current is westward intensified (e.g., Cochrane 1963). Estimates of the Loop Current transport vary from 22-33 Sv (e.g., Ochoa et al. 2001 and Sheinbaum et al. 2002 and references cited therein). There can be flows southward out of the Gulf at the western and eastern edges of Yucatan Channel (e.g., Cochrane 1963; Hansen and Molinari 1979; Ochoa et al. 2001; Sheinbaum et al. 2002). Ochoa et al. (2001) estimated southward transports of 8 Sv.

As it enters the Gulf, the Yucatan Current extends to the sill depth of approximately 2000 m. Its continuation in the Gulf, which is called the Loop Current, carries within it water masses from the global ocean. The Loop Current extends into the eastern Gulf and then exits through the Florida Straits. The net transport out of the Florida Straits has been estimated at approximately 32 Sv (Hofmann and Worley 1986) upstream of the sill and 32 ± 3 Sv downstream of the sill (Schmitz and Richardson 1968; Baringer and Larsen 2001).

36

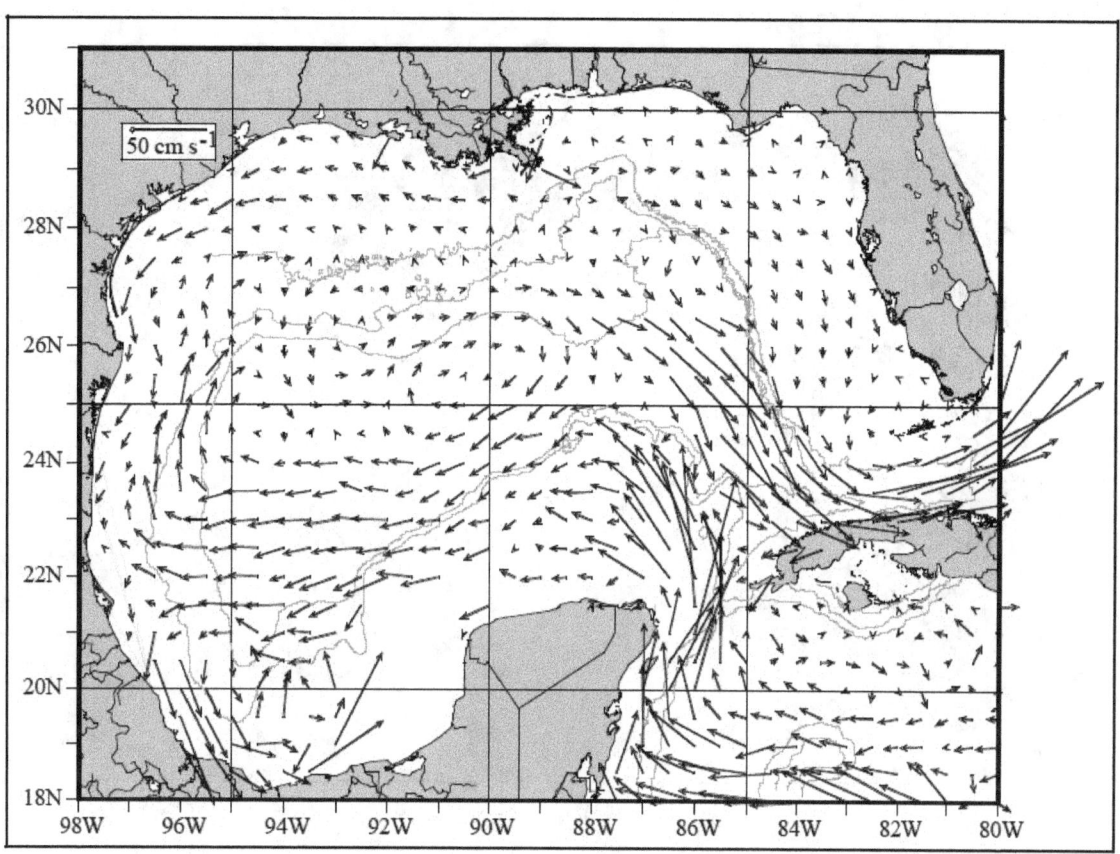

Figure 4.2.   Near-surface drifter velocity estimates. Estimates were made for each 0.5° x 0.5° bin based on averaging all drifter velocity estimates in that bin for the period 1989-1999. Shown are the 200-, 1000-, and 3000-m isobaths. (from Nowlin et al. 2001)

As it extends into the Gulf, the Loop Current may be confined to the southeastern Gulf (Figure 4.4, lower panel) or may extend well into the northeastern or north central Gulf (Figure 4.4, upper panel). It can interact with the northern slope (see e.g., Huh et al. 1981; Paluszkiewicz et al. 1983). Bunge et al. 2002 showed that transports below ~1000 m in the Yucatan Channel were well correlated to the surface extension of the Loop Current and suggested this relationship results in deep exchange of Caribbean and Gulf waters. Fratantoni et al. (1998) concluded that Loop Current perturbations affect the formation and evolution of Tortugas eddies in the southern Straits of Florida. Extensions of the Loop Current into the northern Gulf can result in the separation of an anticyclonically circulating Loop Current Eddy (LCE) that then may migrate into the western Gulf, if it does not reattach to the Loop current (e.g., Maul 1976; Vukovich 1995; Sturges and Leben 2000).

Currents associated with the Loop Current and its eddies extend down to at least the sill depth of the Florida Straits (somewhat less than 800 m) and decrease with depth. Weatherly (2004) found that currents in the Loop Current and its LCEs extended down to a depth of at least 900 m. Using data from 1972 and 1973 to determine mass and salt balances, Maul (1976) estimated that approximately 10% of the water above 700 m that entered through the Yucatan Channel was exchanged with the waters of the Gulf before flowing out the Florida Straits. His results likely were influenced by an LCE separation event that occurred during the measurements.

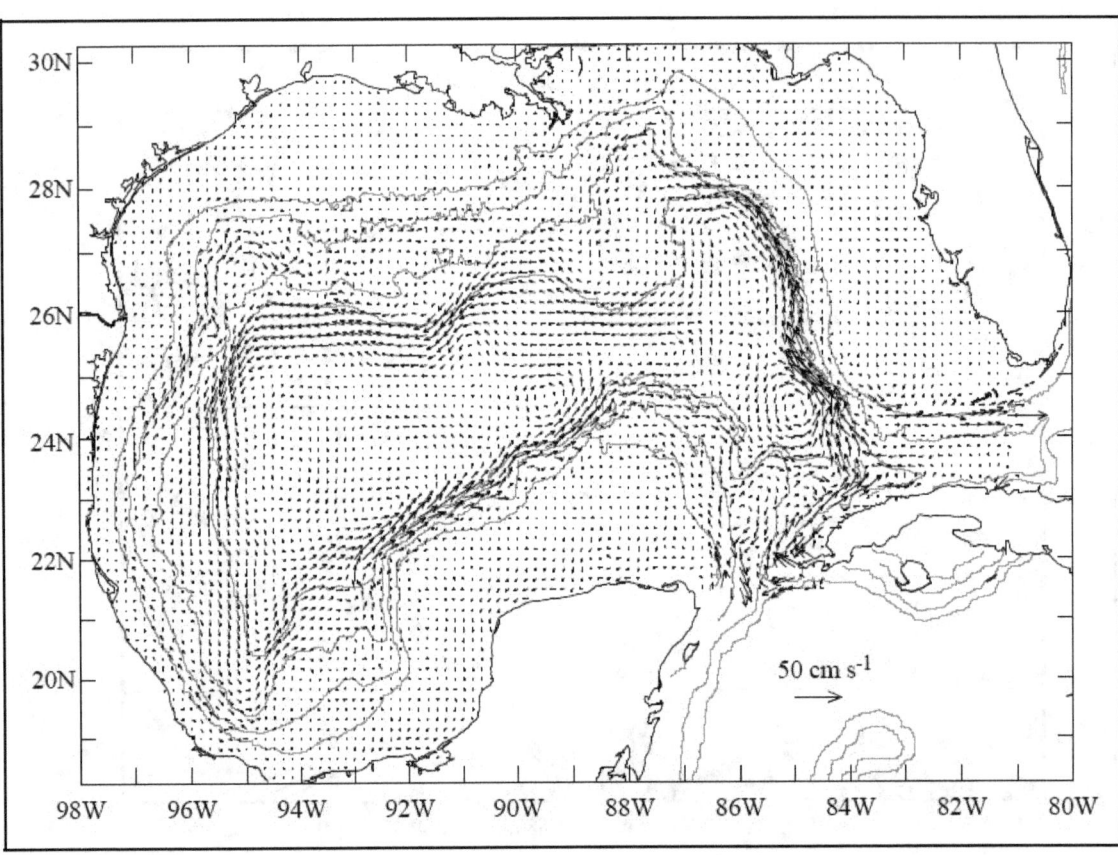

Figure 4.3.   Record-length mean current vectors at the first sigma level above the bottom. Constructed from 1993-1999 University of Colorado Princeton Ocean Model output. Shown are the 200, 1000, 2000, and 3000-m isobaths. (from Nowlin et al. 2001)

The anticyclonic LCEs separate from the Loop Current at intervals ranging from 3 to 17 months, with a peak in frequency of separation near 12 months and secondary peaks at 9 and 6 months (e.g., see Sturges 1994; Sturges and Leben 2000). The LCEs can remain in the eastern Gulf for some time after their formation and can reattach to the Loop Current (e.g., Vukovich 1995). Most eventually move westward, with typical translation speeds of approximately 5 km·d^{-1} (Elliot 1982, see also Hamilton et al. 1999), and reach the western edge of the Gulf of Mexico basin. The LCEs have current cores and water masses similar in properties to waters of the Caribbean type that comprise most of the Loop Current waters. Thus, their migration westward carries water with these properties into the western Gulf. These anticyclonic, mesoscale current rings have average lifetimes longer than one year (Elliot 1982) and may spawn cyclonic rings during interaction with one another or with the continental slope (e.g., Vidal et al. 1994). Hamilton et al. (2002) reported cyclones moved on and off the slope in association with large anticyclones and that cyclone-anticyclone eddy pairs were a mechanism for exchange of material between the shelf and deep waters (see also Hamilton 1992 on cyclonic eddies). The migration of the LCEs and other eddies represent one mechanism for transport of Loop Current waters throughout the Gulf of Mexico.

The second forcing function is wind stress, which primarily affects the upper waters of the Gulf of Mexico. This forcing generates two main classes of phenomena. First, low frequency regional wind patterns can generate low frequency regional circulation patterns. Sturges (1993) suggests a wind-

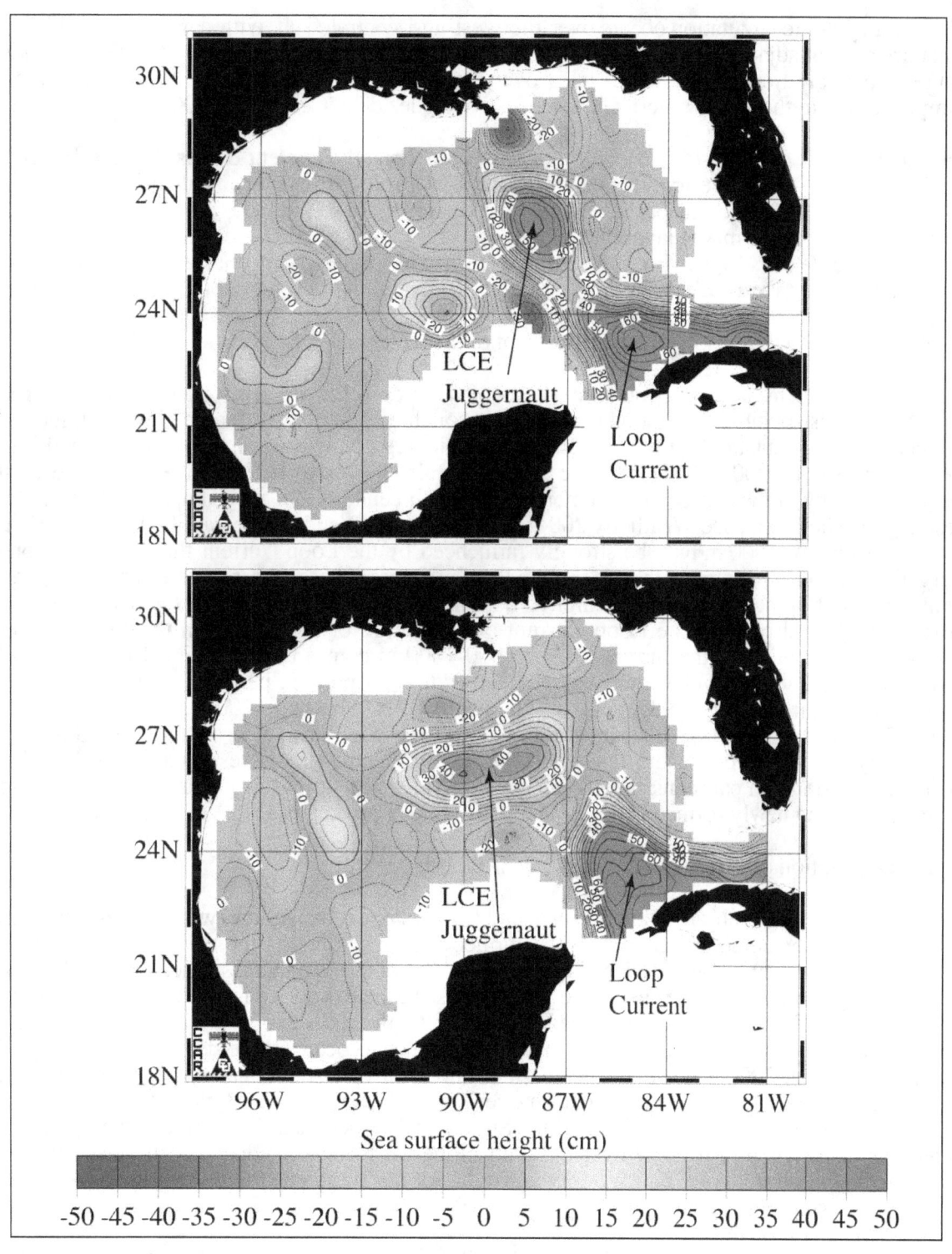

Figure 4.4. Sea surface height fields in the Gulf of Mexico. Fields are for (upper) 20 September 1999 showing extension of Loop Current into the northeastern Gulf and formation of LCE Juggernaut and (lower) 14 November 1999 showing westward migration of Juggernaut, with the Loop Current in the southeastern Gulf, and an older anticyclone in the western Gulf.

driven anticyclonic circulation occurs over the west and central Gulf with a westward intensified boundary current adjacent to the western slope of the Gulf. Vázquez de la Cerda (1993) has shown that seasonal variations in the wind fields over the Gulf generate a cyclonic circulation over the Campeche Bay in the southwestern Gulf (see also Vázquez de la Cerda et al. 2005).

The second class of wind-forced currents consists of episodic current events that are forced by highly energetic atmospheric events such as tropical cyclones, extratropical cyclones, cold air outbreaks, and other frontal passages. The strongest of these includes hurricanes which can produce currents in the mixed layer that exceed 150 cm·s^{-1}, and when combined with wave currents, may exceed 300 cm·s^{-1} (Nowlin et al. 2000). Hurricane induced currents at 200 m have been observed to be approximately 100 cm·s^{-1} in several instances, and even at 700 m may reach speeds of 15-20 cm·s^{-1} (e.g., Brooks 1983). These storms can inject air bubbles deep into the upper water column, enhancing the dissolved oxygen content of the water.

Below about 800 m, the direction and speeds of deep currents are not well understood. Model output and geostrophic calculations based on reasonable choices of reference levels suggest the long-term mean circulation of the deep Gulf is cyclonic (e.g., Nowlin et al. 2001; Welsh and Inoue 2002; Sturges et al. 2004). This cyclonic circulation also seems to be intensified offshore of the steep topography of the Sigsbee Escarpment in the north central Gulf, the Campeche shelf, and the west Florida shelf (e.g., see Weatherly 2004). Hamilton (1990) and Sturges et al. (1993) suggested that currents below 1000 m may be strongly influenced by the Loop Current through excitation of energetic currents associated with topographic Rossby waves. These currents were barotropic, being highly coherent in the vertical, and exhibited bottom intensification. Evidence suggests these deep currents may be driven by the Loop Current or separation of an LCE from the Loop Current. Additional information on the currents beneath 800-1000 m comes from numerical models. Model studies suggest that both cyclonic and anticyclonic eddies are present, perhaps as pairs, in the deep basin waters (Sturges et al. 1993; Welsh and Inoue 2000; Nowlin et al. 2001; see Figure 4.3). The eddies form in the eastern Gulf under the Loop Current or newly formed LCEs. Guided by topography, they then propagate into the western Gulf. Some model results suggest deep layer eddies form, often as a pair consisting of an anticyclone and a stronger cyclone, in a response to a westward moving newly formed anticyclonic LCE in the upper layer (e.g., Welsh and Inoue 2000).

4.1.2   Description of Water Masses in the Gulf of Mexico

Surface waters in the Gulf are subject to heat and freshwater exchanges and winds. These processes may affect the oxygen concentrations in the surface waters by altering the rate at which oxygen is exchanged with the atmosphere, which is the main source of oxygen in the surface waters. Although the buoyancy forcing by river discharge that affects the nearshore coastal currents is a form of thermohaline forcing that is known to be important over the shelves of the Gulf, no thermohaline forcing of consequence or water mass formation are known to occur in the deepwater region of the Gulf (Nowlin et al. 2000). However, Nowlin and Parker (1974) suggested that, during winter time cold air outbreaks, there may be formation over the shelves of a water type with characteristics similar to the waters just below the core of the Subtropical Underwater. They further suggested this water was dense enough to flow over the Gulf slopes at the intermediate depths just below the Subtropical Underwater core. This water type would not reach the deep Gulf, and hence would not be a source of oxygen to the deep waters.

In general then, water masses in the Gulf are the global ocean water masses that enter and exit the Gulf through the Yucatan Channel and the Florida Straits. The sill depths of these bathymetric features are approximately 2000 and 800 m, respectively. The sills essentially control which water masses enter and leave the Gulf. Source waters for oxygen in the deep part of the water column mainly originate in the Atlantic Ocean (Morrison and Nowlin 1982). They are modified as they spread from the Caribbean Sea into the Gulf.

There are five water masses carried into the Gulf by the Loop Current (Table 4.1). These are described briefly from top to bottom. The Subtropical Underwater is characterized by a salinity maximum at its core. It occurs at typical depths of 150-250 m in the eastern Gulf (Morrison and Nowlin 1977) and, if present, between 0-250 m in the western Gulf (Morrison et al. 1983). Below the Subtropical Underwater is 18°C Sargasso Sea water, which is characterized by an oxygen maximum between 3.6-3.8 mL·L^{-1} typically at depths of 200-400 m in the Loop Current waters in the eastern Gulf (Morrison and Nowlin 1977). Within the general decrease of oxygen with depth that occurs in the upper 1000 m, the presence of the oxygen maximum of the 18°C Sargasso Sea water together with any shallow sub-surface local maximum creates a double oxygen minimum structure in the eastern Gulf (Shiller 1999). This oxygen maximum appears to erode away as the waters mix into the western Gulf (Morrison et al. 1983).

The oxygen minimum layer in the Gulf of Mexico develops outside the Gulf and enters through the Yucatan Channel as an established water mass feature (Wüst 1964). This water mass is the Tropical Atlantic Central Water, which is characterized by an oxygen minimum of below 3 mL·L^{-1}. It occurs at depths of 450-700 m in the eastern Gulf (Morrison and Nowlin 1977) and 250-500 m in the western Gulf (Morrison et al. 1983). The oxygen minimum layer spreads across the Gulf of Mexico and can intersect the seabed on the continental slope (Kennicutt 2000). Figure 4.5 shows data collected in May/June 2000 during a cruise of the MMS-sponsored Northern Gulf of Mexico Continental Slope Habitats and Benthic Ecology (DGOMB) Study. The Tropical Atlantic Central Water is seen as the oxygen minimum zone centered about the 27.15 sigma-theta density surface. These recent data are consistent with findings of the earlier studies.

Other water masses in the deep water Gulf include the Antarctic Intermediate Water, which occurs between 700-1000 m depth in the eastern Gulf and 500-1100 m in the western Gulf, and a modified upper North Atlantic Deep Water, which is between 900-1200 m in the eastern Gulf and 1000-1100 m in the western Gulf (Nowlin and McLellan 1967; Morrison and Nowlin 1977; Morrison et al. 1983). These deeper water masses do not have oxygen as their chief characteristic water property.

Nowlin et al. (1969) found that, from approximately 1500 m to the bottom, dissolved oxygen distributions suggest no clearly discernible horizontal variation and exhibit only slight vertical gradients. They found mean concentrations of approximately 5 mL·L^{-1}. Temperature and salinity below the sill depth exhibit similar distributions (McLellan and Nowlin 1963; Nowlin and McLellan 1967). Data from the early 1960s and the 1970s, as well as data collected during DGOMB, indicate that the deep water of the Gulf of Mexico is homogeneous with relatively constant dissolved oxygen content (see Figure 4.5); this observation is examined further in Section 4.3.

The deep water masses of the Gulf mainly enter with the Loop Current through the Yucatan Channel and then flow into the Gulf interior to fill the deep basins. Using silica as a tracer, Carder et al. (1977) found a bolus of renewal waters, 200-300 m thick, moved from the sill region into the Gulf, sank rapidly to the bottom at 3500 m with some mixing, and penetrated at least 350 km into the Gulf. To explain observed temperature-salinity distributions, Maul et al. (1985) suggested upper North Atlantic Deep Water flows into the Gulf on the east side of the Yucatan Channel, cyclonically flows around the Gulf basin, and returns to the channel to flow southward on the west side. Sturges et al. (2004) used temperature to infer cyclonic circulation at the 2000-m depth. Model results of Welsh and Inoue (2002) seem to support this idea. They propose a conceptual model for ventilation of the deep waters. The relatively cold, salty, oxygen-rich waters that enter the Gulf at depth through the Yucatan Channel are downwelled in the eastern basin and then circulated cyclonically into the rest of the Gulf. Older, warmer water is upwelled onto the Campeche Bank beneath the western limb of the Loop Current.

Table 4.1

Water Masses of the Caribbean Sea and Gulf of Mexico

Water Mass	Depth (m)	σ_θ (kg·m^{-3})	Characteristic Feature	Characteristic Range
Eastern Caribbean[1]				
SUW-LC	150-250	25.40	S_{max}	
SUW				
18°C W	200-400	26.50	$O_{2\ max}$	
TACW	400-700	27.15	$O_{2\ min}$	
AAIW	600-800	27.30	$NO_{3\ max}$	
AAIW	700-900	27.40	$PO_{4\ max}$	
AAIW	600-900	27.40	S_{min}	
AAIW	800-1000	27.50	$SiO_{3\ max}$	
	1100-1600	27.70	S_{max}	
UNADW	at sill	27.75	$SiO_{3\ min}$	
	at sill	27.75	$PO_{4\ min}$	
	at sill	27.75	$NO_{3\ min}$	
Eastern Gulf of Mexico[2]				
SUW-LC	150-250	25.40	S_{max}	36.7-36.8
SUW	150-250	25.40	S_{max}	36.4-36.5
18°C W	200-400	26.50	$O_{2\ max}$	3.6-3.8 ml·l^{-1}
TACW	400-700	27.15	$O_{2\ min}$	2.85-3.25 ml·l^{-1}
AAIW	na	na	$NO_{3\ max}$	na
AAIW	700-900	27.40	$PO_{4\ max}$	1.8-2.5 mg-at·l^{-1}
AAIW	800-1000	27.50	S_{min}	34.86-34.89
AAIW		27.50	$SiO_{3\ max}$	*
UNADW				
MIX**	900-1200	27.70	$SiO_{3\ max}$	23-25 mg-at·l^{-1}
Western Gulf of Mexico[3]				
SUW-LC	na	na	na	na
SUW	0-250	25.40	S_{max}	36.4-36.5
18°C W	na	na	na	na
TACW	250-400	27.15	$O_{2\ min}$	2.5-2.9 ml·l^{-1}
AAIW	500-700	27.30	$NO_{3\ max}$	29-35 mg-at·l^{-1}
AAIW	600-800	27.40	$PO_{4\ max}$	1.7-2.5 mg-at·l^{-1}
AAIW	700-800	27.50	S_{min}	34.88-34.89
AAIW				
UNADW				
MIX**	na	na	na	na

1 - Morrison and Nowlin 1982   2 - Morrison and Nowlin 1977; Nowlin and McLellan 1967
3 - Morrison et al. 1983; Nowlin and McLellan 1967
SUW-LC = Subtropical Underwater within the Loop Current and Caribbean Sea
SUW = Subtropical Underwater in the Gulf but outside the Loop Current
18°C W = 18°C Sargasso Sea Water;   TACW = Tropical Atlantic Central Water
AAIW = Antarctic Intermediate Water;   UNADW = Upper North Atlantic Deep Water
**MIX = Mixture of low silicate UNADW and very high silicate Caribbean Mid-Water
*high SiO_3 in AAIW and MIX waters results in broad SiO_3 maximum approximately from 27.50 to 27.70

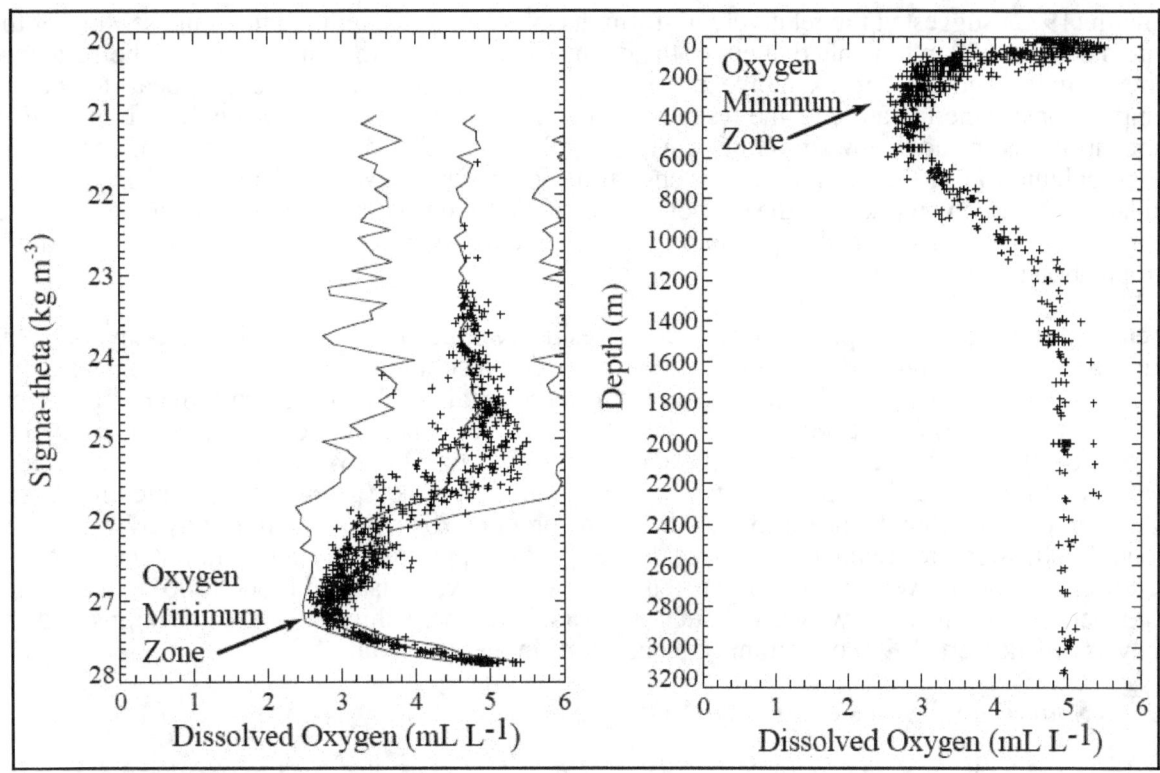

Figure 4.5. Dissolved oxygen profiles for DGOMB cruise 1 in May/June 2000. Shown are profiles in density (left) and depth (right). Lines in the left panel are the mean and ±2.3 standard deviations based on a combination of data from several recent cruises (DGOMB cruise 1, ten LATEX A cruises, and nine NEGOM-COH cruises).

4.1.3   Flushing Time

The flushing time, also called residence time, is the time required to replace an amount of a substance (or volume of water) equal to the amount stored in the reservoir (e.g., Broecker and Peng 1982). It can be expressed as

$$\tau = M/R$$

where τ is the flushing time, M is the mass or the amount of a substance in the reservoir, and R is the flux or the rate of supply or removal of the substance to or from the reservoir (e.g., Libes 1992; Pilson 1998). Two major assumptions in this calculation are that the reservoir is well mixed and the concentration of the substance in the reservoir does not change.

The rate of supply of oxygen to the deep Gulf depends on the flux of oxygen both into and out of the various depth zones of the Gulf. This flux depends on the concentration of dissolved oxygen in the source waters, the transport of the source waters into the Gulf interior, and the rate of removal from respiration and detrital decay/remineralization. The concentrations in the source waters are relatively well known for the various depths, but the transports and consumptive rates within depth layers are not.

43

Nowlin (1972) suggested the relatively uniform dissolved oxygen, temperature, and salinity of the waters below 1500 m indicate that either the deep waters have common sources or the residence time is great enough for exchange processes to erode gradients. There appear to be no computations in the literature of the residence time of oxygen or of deep waters based on oxygen concentrations in the deepwater Gulf of Mexico. However, El-Sayed et al. (1972) made a very rough calculation of the minimum residence time for waters between 100 and 750 m based on oxidation of carbon produced in the euphotic zone. They found the minimum residence time was 3 years, but that the actual residence time must be much longer because of the simplicity of the computation made.

There have been several papers that considered residence times for water in the deep layers of the Gulf based on other properties. Shiller (1999) provided a summary of the literature on residence times of deep Gulf waters. Using silica data from Carder et al. (1977) for the eastern Gulf and from Tréguer et al. (1995) for the world ocean, he estimated a residence time of 50 years for deep Gulf waters. Shiller (1999) reported that the carbon-14 data of Mathews et al. (1973) indicated a residence time of 270 years in the western Gulf. Schink (private communication) used radiochemistry to estimate the turnover to be on the order of 70 yr. A modeling study by Welsh and Inoue (2000) suggested relatively short residence times of approximately 10 years for the waters of the lower layer. However, in their more recent modeling study, Welsh and Inoue (2002) suggest the residence time of the deep water is about 100 years. They base this on flux calculations and the analysis of Buerkert (1997) of barium concentrations in the deep Gulf.

4.2   Sources and Sinks of Dissolved Oxygen

As described in the introduction to this chapter, the distribution of dissolved oxygen in the ocean is dependent on the complex interplay between biological, chemical, and physical processes. Because it is a biologically active gas, dissolved oxygen is a non-conservative property of sea water. In order to develop a model that describes the vertical and horizontal patterns of oxygen in the deepwater Gulf of Mexico, the major sources and sinks of oxygen were identified by reviewing available literature (e.g., Riley 1951, Shiller 1999, and Kennicutt 2000 and references therein). Dissolved oxygen in the deepwater Gulf of Mexico can be traced to global ocean water masses, to physical processes such as advection and mixing, as described in Section 4.1, and to consumption through biogeochemical processes in the water column and in surficial sediments. The sources and sinks of oxygen are summarized in Table 4.2, together with their vertical spheres of influence.

For purposes of this study, the shelf is defined to be that part of the Gulf in water depths of 200 m or less. The wide continental shelves of the Gulf tend to isolate the deepwater Gulf of Mexico from many coastal processes (Kennicutt 2000) that might otherwise be sources or sinks for oxygen. This includes the direct effect of river discharges on air-sea interactions and river-oceanic water mixing. No water masses of consequence are formed over the shelf. Discharges from the Mississippi River have significant impact on the oxygen concentrations in the water over the Louisiana shelf, both from the increased water stratification caused by the buoyancy effects of the fresh water and from the increased nutrients carried by riverine water into the ocean. Such nutrients then become available to fuel an increase in organic matter production that contributes to oxygen consumption, and even hypoxia, over the shelves. However, these effects are limited mainly to the inner shelf inshore of water depths less than about 60 m (e.g., Rabalais and Turner 2001).

Under some circulation regimes, sediments discharged from the Mississippi River can increase the quantities of particulate matter in the open ocean over the slope in the central Gulf. As the organic matter in these particles sinks and decays, the oxygen consumption in this region is expected to increase. However, little of the Mississippi River plume is transported into the deep Gulf. Shiller (1999) examined dissolved oxygen concentrations, averaged over 2° squares, at the depth of the oxygen minimum. He suggested that proximity to continental margins may affect oxygen concentrations in intermediate-depth waters. This would result from an increase in oxygen consumption processes associated with the more productive margins as compared to the less productive open ocean. The effects of shelf processes on oxygen concentrations are examined in Section 4.3.

Table 4.2

Sources and Sinks of Dissolved Oxygen in the Deepwater Gulf of Mexico

Vertical Zone	Process Description
Euphotic Zone	Transport of water masses
	Horizontal advection
	Vertical mixing
	Diffusion
	Exchange with the atmosphere (air-sea interface)
	Oxygen production by primary productivity (photosynthesis)
	Aerobic microbial respiration
	Macrofauna, meiofauna respiration
	Anaerobic bacterial respiration in microenvironments (particles)
	Organic matter oxidation
	POC, DOC, CO_2, reduced species (H_2S, Fe^{2+}, Mn^{2+}, NH^{4+}) rain rate
Below the Euphotic Zone	Transport of water masses
	Horizontal advection
	Vertical mixing
	Diffusion
	Aerobic microbial respiration
	Macrofauna, meiofauna respiration
	Fishes, zooplankton respiration
	Anaerobic bacterial respiration in microenvironments (particles)
	Organic matter oxidation
	POC, DOC, CO_2, Reduced species (H_2S, Fe^{2+}, Mn^{2+}, NH^{4+}) rain rate
Water/Sediment Interface	Sediment-water diffusive flux
	Water column macrofauna, meiofauna respiration
	Water column aerobic microbial respiration
	Water column anaerobic bacterial respiration in microenvironments (particles)
	Benthic oxygen demand
	Epifauna respiration
	Sediment aerobic microbial respiration
	Sediment macrofauna, meiofauna respiration
	Sediment anaerobic microbial respiration (in the oxygen free zone)
	POC burial
Other	Oil seeps
	Oil spills
	Petroleum extraction (produced waters, drilling fluids, drill cuttings)
	Other anthropogenic processes

There are three basic regimes with differing physical-biogeochemical processes that affect dissolved oxygen concentrations in the water column (Figure 4.6). First is the surface layer, which encompasses the euphotic zone, the mixed layer, and the interface with the atmosphere. In the deep ocean, it typically extends from the sea surface down to ~100 to 200-m depth. In this zone, exchanges with the atmosphere add or remove oxygen from the surface water and wind events mix well-oxygenated surface waters to the bottom of the mixed layer. Photosynthesis adds oxygen locally in this zone while respiration from both animals and particulate decay processes removes it. The second regime extends from the bottom of the surface layer to just above the water-sediment interface. It consists of three components defined by the sill depths of the Yucatan Channel (~2000 m) and the Florida Straits (~800 m). This part of the water column includes the main thermocline and the deep waters. Here oxygen generally is removed through respiration and particulate decay processes and the oxygen minimum zone is found. The third regime is the water-sediment interface at the sea floor. It too is dominated by respiration and particulate decay processes that remove oxygen. Physical processes associated with advection and vertical mixing of water masses, and, to a comparatively negligible extent, with diffusion import or export oxygen in all three regimes. In some cases, water movement can offset the results of local processes of photosynthesis and respiration. Below are summarized the essentials of the processes that affect dissolved oxygen concentrations and distributions in the oceans.

4.2.1 Physical Processes in the Gulf of Mexico

The basic control on dissolved oxygen concentrations in near-surface waters is exchange with the atmosphere. The flux of oxygen between the atmosphere and the ocean depends on wind speed, air and sea surface temperature, salinity, surface films, wave action, and bubble injection rates (e.g., Richards 1965; Riley and Chester 1971; Broecker and Peng 1982; Pilson 1998). In some cases, waters within the photic zone can become supersaturated with respect to oxygen. This is caused by air injection or photosynthesis (e.g., Broecker and Peng 1982). The exchange occurs across a thin layer of the atmosphere-surface water interface. Near surface waters usually are at or near equilibrium concentrations.

Oxygen saturation levels depend on the relative partial pressures of the air and water media and, hence, on factors that control gaseous solubility. Solubility is a function of temperature, salinity, barometric pressure, and atmospheric humidity. Of these, the temperature and salinity of the water are the principal factors. Humidity may be negligible in controlling solubility. Solubility decreases with increases in temperature or salinity and increases with increases in pressure. Other factors that influence oxygen concentrations include surface conditions (turbulence), presence and extent of surface organic or inorganic films, kinetics of dissolution, and surface heating (Riley and Chester 1971). Oil spills would change the structure of the surface films, and would generally decrease the solubility of the water and thereby decrease air-sea exchanges of oxygen (NRC 2003).

At the air-sea interface, by definition equilibrium implies that the exchange of oxygen molecules in both directions is equal; it does not imply the process is stagnant. The rate of exchange between the gaseous and liquid phases is controlled by the rate of molecular diffusion through two boundary layers, one in the gaseous phase and one in the liquid phase. Molecular diffusion in the gaseous phase is several thousand times greater than in the liquid phase. Thus, when the boundary layer thickness in the water decreases with increasing turbulence, the exchange rates between the two media are greatly increased. Note also that wave action can result in the injection of bubbles deep into the upper water column, increasing oxygen concentrations in the water column due to increased surface area and oxygen undersaturation at depth. When these waters return to the surface they can be supersaturated.

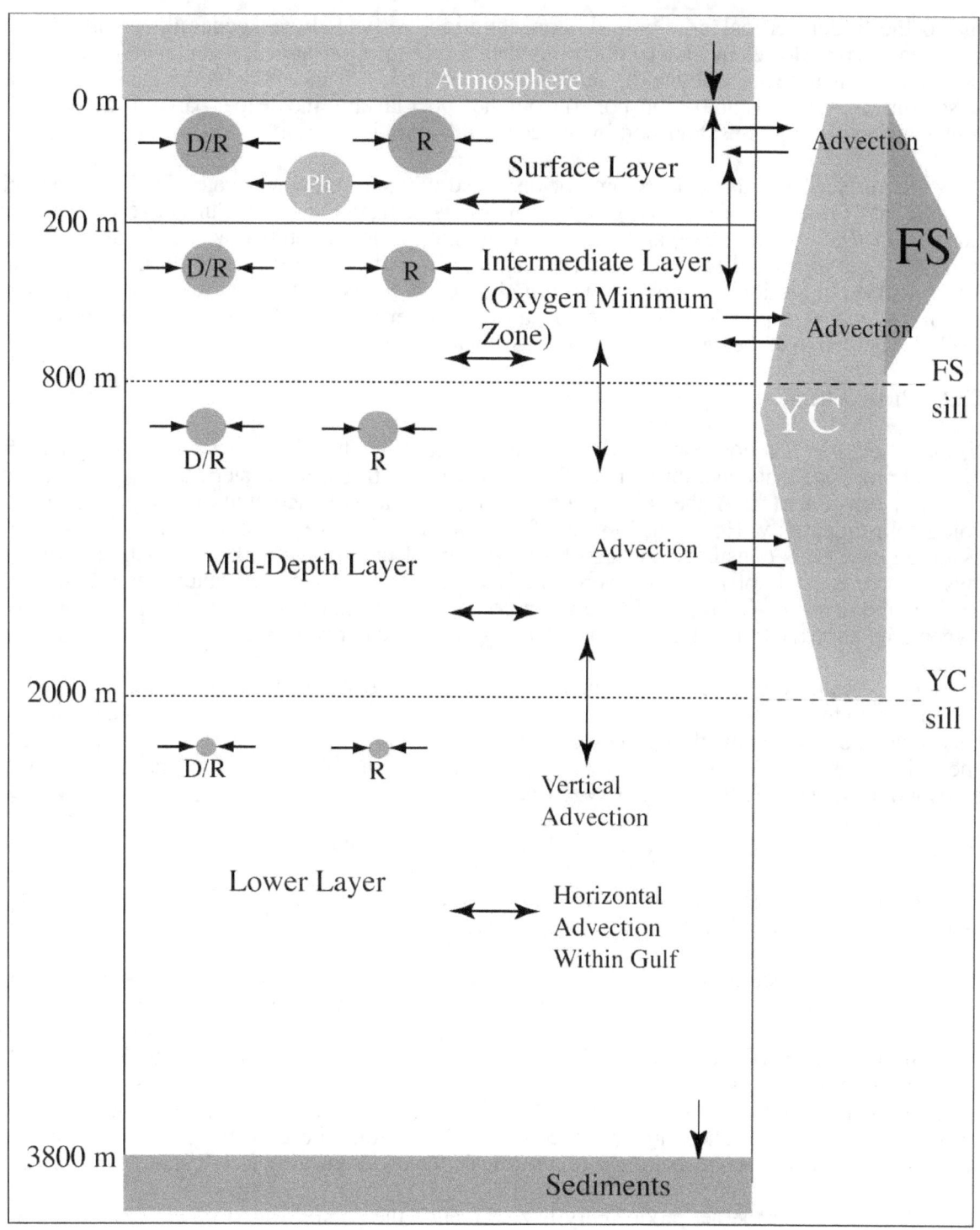

Figure 4.6.   Schematic of vertical zones in the Gulf of Mexico. Inflow and outflow from the Yucatan Channel (YC) and the Florida Straits (FS) are shown on the right with their sill depths. Advection into and within each box and the exchanges at the air-sea can be either direction depending on conditions. Exchanges at the water-sediment interfaces generally are consumptive. Decay of detritus and remineralization (D/R) and respiration (R) are consumptive processes with rates that decay exponentially with depth. Photosynthesis (Ph) is a source.

Due to the larger vertical gradients in concentration, vertical fluxes generally greatly exceed horizontal fluxes. However, under conditions that usually prevail at sea, turbulence produced by wind ensures that the layer above the seasonal thermocline is in thermodynamic equilibrium with the sea surface. Oxygen can be transported from the upper layer to the deeper parts of the ocean by vertical advective processes; transport by molecular diffusion is negligible.

Below the surface layer, the major source of dissolved oxygen in the deepwater Gulf of Mexico is the horizontal advection of well-oxygenated water masses from the Loop Current (e.g., Morrison and Nowlin 1977; Shiller 1999) and vertical and horizontal advection below the Yucatan Channel sill depth (e.g., Welsh and Inoue 2002). The distribution and characteristics of these water masses were discussed in Section 4.1 above. In Section 4.3 below, typical values of oxygen concentrations seen at the various depths are discussed. Because no water masses of consequence are formed in the Gulf, ventilation of the deep waters from thermohaline circulation does not occur.

4.2.2   Biogeochemical Processes in the Gulf of Mexico

Dissolved oxygen is a non-conservative constituent of water that is affected by biological and chemical processes that occur throughout the water column. In surface waters, the major source of dissolved oxygen is from the atmosphere, with net local enhancements from photosynthetic biological productivity. However, in subsurface waters, which are the main focus of this study, dissolved oxygen is consumed through biogeochemical processes, with replenishment due to physical processes of horizontal and vertical advection and mixing of the water masses. Rates of consumption of dissolved oxygen in the water column can be estimated from changes in apparent oxygen utilization (AOU) with changing water mass structure in deep waters.

Biologically induced changes predominantly occur after the water mass is no longer in direct exchange with the atmosphere. The main causes of *in situ* changes in the oxygen concentrations of the sea water are the competing processes of photosynthesis and respiration. In the upper waters, where light is sufficient, photosynthesis by phytoplankton usually predominates. It leads to the removal of CO_2 and the liberation of oxygen, according to the formula

$$6CO_2 + 6H_2O = (C_6H_2O_6) + 6O_2,$$

where the left side represents uptake and the right side release (Chester 2003). Surface values of dissolved oxygen in the Gulf of Mexico are presented in Section 4.3.

Photosynthetic production of oxygen can be a significant source of oxygen (e.g., see Craig and Hayward 1987 in a study in the North Pacific Ocean). Indeed, in regions of intense biological activity, this can result in the development of a water layer supersaturated with dissolved oxygen. Productivity seaward of the shelf edge is generally low in the Gulf (see Biggs and Ressler 2001, and references therein), so it is not expected that there would be wide-spread supersaturation associated with biological activity in the Gulf. However, Biggs and Ressler (2001) identified "hot spots" of primary production along the continental margin. These were localized, sometimes temporally persistent, areas associated with mesoscale features such as cyclonic/anticyclonic eddies.

Below the point where the net productivity is zero (called the compensation point or compensation depth), the dissolved oxygen is only consumed by the respiration of decaying plant materials, animals, and bacteria. Respiration is the reverse of the above chemical equation. Biological respiration can be aerobic or anaerobic in microenvironments. Anaerobic respiration occurs in areas that are oxygen deficient, leading to denitrification or sulfate reduction processes that utilize nitrate and sulfate instead of dissolved oxygen as the electron acceptors. Iron and manganese oxides can also serve as electron acceptors. The present circulation system in the deep Gulf results in well-oxygenated deep waters, so anaerobic respiration is limited to microenvironments (i.e., inside particles) or small stagnant water pockets near the sea floor. Oxygen is consumed by the heterotrophic activity of organisms (primarily bacteria) utilizing organic matter such as particulate organic carbon (POC) and dissolved organic carbon (DOC) in the water column, as well as hydrocarbons and associated gases that enter the water from seeps along the northern continental slope (e.g., MacDonald 2002; De Beukelaer et al. 2003). The ultimate limiting factor in oxygen consumption is the amount of metabolizable organic carbon available.

Oxygen is involved in the remineralization of organic matter to its constitutive elements. The standard representation of remineralization is represented by the Redfield equations. The Redfield ratio was formulated as a $P:N:C:O_2$ of 1:16:106:138; the $P:O_2$ ratio of 1:138 was estimated using assumed ratios of $C:O_2$ of 1:1 and $N:O_2$ of 1:2 (Redfield et al. 1963). (But see Anderson et al. 1986, Anderson and Sarmiento 1994, Anderson 1995, and Hedges et al. 2002, which suggest the Redfield ratios for O_2 may be in error.) Takahashi et al. (1985) revised these estimates for thermocline waters of the Atlantic and Indian oceans and determined them to be 1:16:122($\pm$18):172. As the organic material sinks down in the water column, it is oxidized and consumes the oxygen approximately in proportion to the ratios expressed by the Redfield ratios. Thus, the oxidation of carbon and nitrogen by oxygen can be represented, for example, by

$$CH_2O + O_2 = CO_2 + H_2O$$

$$NH_3 + 2O_2 = HNO_3 + H_2O$$

The rate of oxygen consumption generally decreases with water depth because this rate depends primarily on the concentration of metabolizable organic matter, which in turn decreases with increasing depth below the thermocline. Thus, consumption in deep waters is slow because little organic matter suitable for consumption reaches these depths. The organic matter that reaches deep waters has been substantially reworked as it sinks through the long water column, leaving a relatively oxidation-resistant residue (e.g., Hedges et al. 2002).

The Yucatan channel sill depth of approximately 2000 m acts as a barrier to the very cold waters of 2-3 °C that exist in the Atlantic at depths of about 2-3.8 km. Therefore, water temperatures in the deep Gulf below the sill depth, which are 4-5°C, are warmer than water temperatures at similar depths in the Atlantic. This likely increases respiration rates in the deep Gulf as compared to similar depths in the Atlantic (Rowe et al. 2003).

Oxygen consumption rates for the Gulf apparently have not been calculated. However, Riley (1951) estimated a rate of oxygen consumption in the water column from an evaluation of oxygen distributions for the Atlantic ocean between 54°N and 45°N (Table 4.3). Jenkins (1982) used helium tracers to determine oxygen utilization rates in a region in the eastern subtropical North Atlantic in the upper 1000 m, but below the euphotic zone. He compared his results to those of Riley (1951) and concluded Riley's estimates for the upper waters were accurate to within 10-20%, but that Riley's deeper water values were suspect. Jenkins (1998) used new data from the eastern subtropical North Atlantic and new analysis techniques to determine oxygen utilization rates in the upper 600 m. His results showed that oxygen consumption decreased exponentially with depth. Pilson (1998) estimated the vertical distribution of oxygen consumption for the world ocean based on extrapolations from sediment trap data and concluded these estimates were in good agreement with those of Riley (1951).

Although these rates of consumption are rough estimates and may be underestimates, they point to two facts of importance for the Gulf of Mexico. First, the rate of consumption decreases almost exponentially with depth. Second, even the oxygen in the deep sea will be used up given enough time, unless it is replaced. Due to the rapid resupply of oxygen by advection of the oxygen-rich water masses entering the Gulf, consumptive processes do not exhaust deep-water dissolved oxygen. But without such ventilation the deep waters would soon become stripped of oxygen, as occurs in places such as the Black Sea.

Table 4.3

Estimated Rate of Dissolved Oxygen Consumption in the Atlantic Ocean Between 45°N and 54°N
(from Riley 1951)

Mean depth (m)	O_2 consumption (μmol kg^{-1} yr^{-1})
200	9.1
280	3.5
370	2.2
510	2.4
700	1.5
1000	0.6
1250	0.2
1500	0.07
2000	0.06
2500-4000	0.006

Processes in the water column where dissolved oxygen reacts with dissolved inorganic matter, including sulfides, Fe^{2+}, and Mn^{2+}, also consume oxygen. Oxygen consumption from these processes are not well understood and likely are negligible in the water column (e.g., Rowe 2005).

The oxygen minimum typically observed in vertical profiles is created when the rate of oxygen production is exceeded by the rate of oxygen consumption (e.g., Riley 1951). This minimum develops in regions of high primary productivity. The biological production in the deepwater Gulf, however, is not high enough that the input of sinking and decaying organic detritus reduces the oxygen to low levels in that part of the basin. Instead, the oxygen minimum layer is transported horizontally into the Gulf as part of the water masses entering with the Loop Current (Metcalf 1976; Morrison and Nowlin 1977; Wüst 1964). The dynamics of biological production and oxygen consumption within the Gulf then contribute to maintenance of the oxygen minimum. Where low oxygen waters intersect sediments on the slope, there is potential for localized reduction of oxygen to suboxic levels, depending on the amount of metabolizable organic material available. An oil spill would contribute to localized decreases in oxygen concentrations (see Section 4.2.3 below; e.g., Carney 2001). The oxygen minimum in the Gulf exhibits concentrations of 2.5-2.9 mL·L^{-1} in the western Gulf (Morrison et al. 1983). These concentrations are quite high compared to oxygen minimum zones in other parts of the world, where concentrations drop below 0.5 mL·L^{-1} (e.g., Rogers 2000; Gage et al. 2000).

Near the seafloor, the same respiration reactions occur as in the water column, but at significantly accelerated rates due to the general presence of a benthic nepheloid layer rich in organic matter (e.g., Karl et al. 1976). Below this layer, oxygen is transported into the sediments by exchange across the interface that is driven by the physical forces of advection and diffusion and the biologic processes of bioturbation and bioirrigation. Oxygen can be consumed in both abiotic and biotically driven redox reactions with reduced components within sediments that may diffuse into the overlying water. These include H_2S, Fe^{2+}, and Mn^{2+}. The net apparent consumption of dissolved oxygen by sediments is commonly referred to as "benthic oxygen demand" (BOD) and biologists commonly equate it to net sediment heterotrophic activity. Results of geochemical studies from DGOMB data found that O_2 concentrations in sediments decreased in a highly variable manner in different parts of the Gulf, reaching undetectable values at depths from <1 mm to ~13 cm below the sediment-water interface (Rowe 2005). This consumption of oxygen by sediments is typical of oceanic sediments. Jahnke and Jackson (1992) estimated the benthic oxygen flux from organic carbon burial rates for the Atlantic and Pacific oceans. Their results estimated the flux in the Gulf and adjacent Atlantic to be $0.06 - 0.24$ mol $O_2 \cdot m^{-2} \cdot yr^{-1}$. The main source of the oxygen to the sediments is from the overlying water. If the transport of dissolved oxygen to these waters is slow or absent, as may occur in sea floor depressions or brine pools, localized pockets of oxygen-depleted waters can be created and anoxic processes can take place in the sediments (see e.g., Pilson 1998).

4.3   Data Reanalysis

For the data analyses below, only bottle data that passed the primary and secondary QA/QC processing were used. For these analyses, 35792 data points from 2790 separate stations were used (see Table 3.1 for a list of the cruises). Figure 4.7 shows the locations of these stations relative to different sub-regions of the Gulf of Mexico.

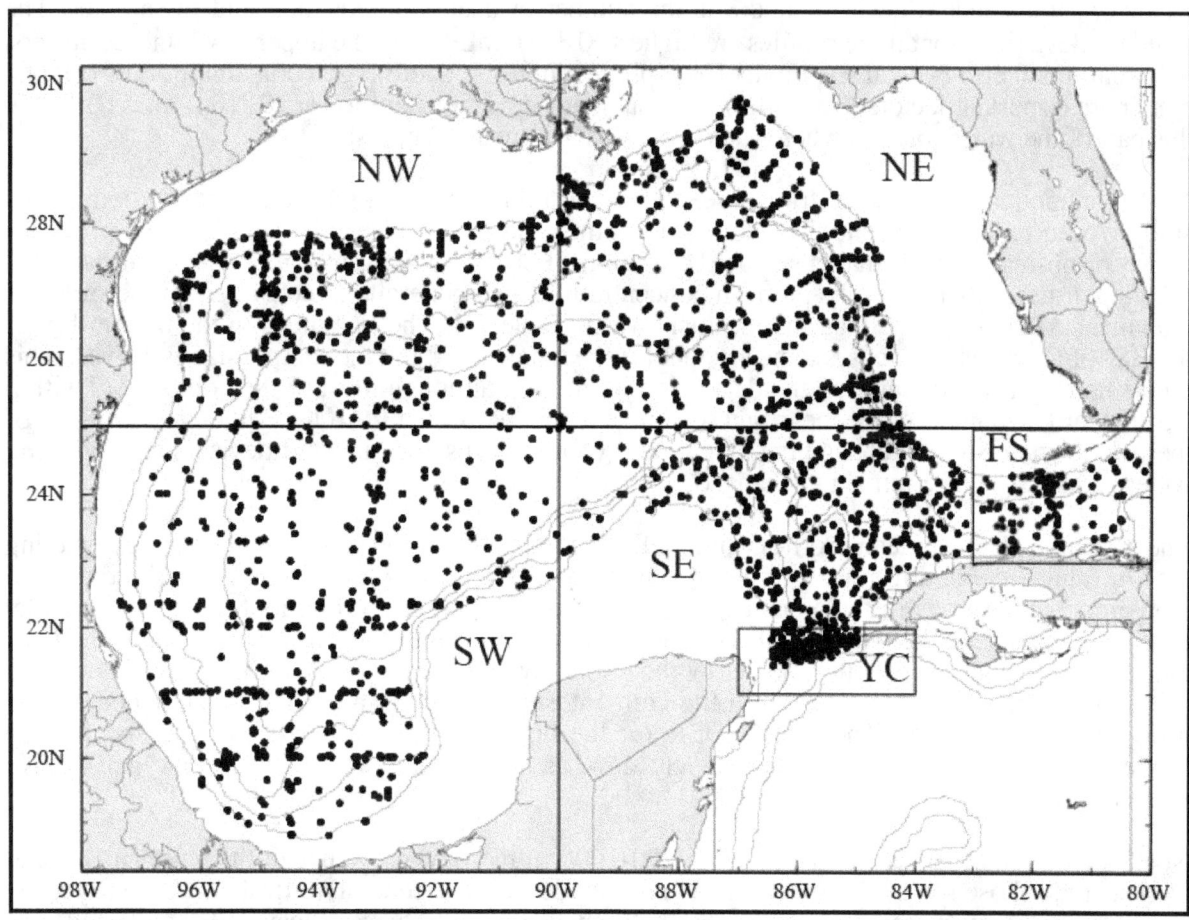

Figure 4.7.   Locations of 2790 stations with at least one good oxygen datum in the Gulf of Mexico. These data were used in the computations of the basic statistics. Geographical regions are shown, including the Yucatan Channel (YC) and the Florida Straits (FS). Bathymetry contours are 200, 1000, 2000 and 3000 m.

51

4.3.1   Basic Statistics

Regional Statistics: The mean and standard deviation were computed by depth bins for the whole Gulf of Mexico as well as for sub-regions within the Gulf. Statistics were calculated for the data that were in the Oxygen Archive as of 22 June 2004 (see Table 3.1).

Figure 4.8 shows the vertical profiles of the mean for the whole Gulf, as well as the Yucatan Channel and Bay of Campeche sub-regions. Table 4.4 provides the statistics. The mean dissolved oxygen concentration profile for the whole Gulf of Mexico follows the classical pattern with high values near surface, decreasing values below this to an oxygen minimum, and generally increasing values with depth below. The profiles for the Yucatan Channel and the Bay of Campeche follow a similar pattern.

Figure 4.8 also shows the standard error of the mean for each bin. The standard error is defined as

$$\sigma_{\bar{x}} = \frac{\sigma}{\sqrt{n}}$$

where σ is the standard deviation and n is the number of data points or independent samples. The standard deviations for all the profiles are highest (0.3-0.6 mL·L^{-1}) in the upper few hundred meters, where atmospheric exchanges, inputs from photosynthesis, and oxygen consumption of organic matter are important factors (Table 4.4). The standard deviations are lower (0.1-0.3 mL·L^{-1}) below this part of the water column, where such processes are greatly diminished.

Note that application of more stringent exclusion criteria to the data set than were applied in this study would result in rejection of more data points and consequently less differences in means between sub-regions and smaller standard deviations. For example, exclusion of data from the 1962 *Hidalgo* cruise, 62-H-3, results in mean concentrations that generally are elevated on the order of 0.01 mL·L^{-1} for many bins than the mean shown in Table 4.4. The low mean in the bottom bin of the Bay of Campeche increases to nearly that for the whole Gulf, but the sample size is halved. This cruise had significant scatter in the data (see Figure 3.7), although most of the points were within 2.3 standard deviations of the mean and so were not rejected in the QA/QC. Although exclusion of these data would slightly change the magnitudes of the means and standard deviations, the general patterns discussed here remain unchanged.

The Yucatan Channel mean profile, which reflects the profile of the source water masses entering the Gulf, has several differences of note from the profile for the whole Gulf. First, the dissolved oxygen concentrations in the Yucatan Channel tend to be higher throughout most of the water column than those of the whole Gulf. Because there are no deep water masses formed in the Gulf and oxygen is consumed with time below the upper layer that interacts with the atmosphere, this difference suggests that the waters in the Gulf interior are not immediately replenished by the inflowing water masses. The mean profile for the upper 750 m in the Bay of Campeche also suggests this replenishment may be relatively slow, as these waters are depleted in oxygen relative to the mean.

Note, however, that between approximately 750-1000 m the oxygen values in the Yucatan Channel are less than those for the whole Gulf. This is evidence of a general vertical rise of isopycnals, carrying their water masses, that occurs in the Gulf because of "mass adjustments associated with the current regime" (Morrison et al. 1983) as the water masses are circulated westward by the currents (see, e.g., Table 4.1). Thus, the more oxygenated waters below 1000 m in the Yucatan Channel rise in depth as they spread into the Gulf giving the observed higher oxygen concentrations just below the oxygen minimum in the waters within the Gulf.

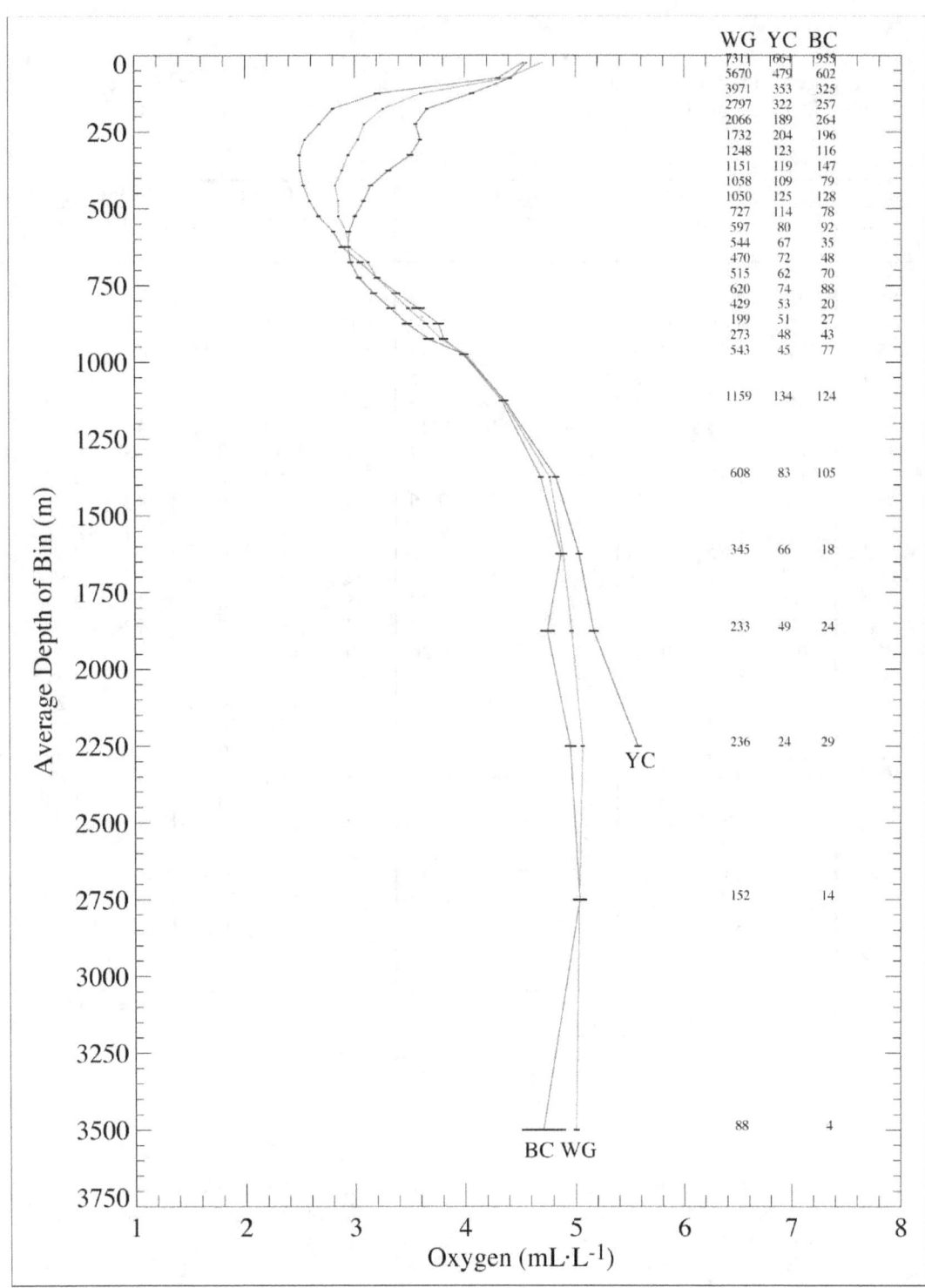

	WG	YC	BC
	7311	1664	1955
	5670	479	602
	3971	353	325
	2797	322	257
	2066	189	264
	1732	204	196
	1248	123	116
	1151	119	147
	1058	109	79
	1050	125	128
	727	114	78
	597	80	92
	544	67	35
	470	72	48
	515	62	70
	620	74	88
	429	53	20
	199	51	27
	273	48	43
	543	45	77
	1159	134	124
	608	83	105
	345	66	18
	233	49	24
	236	24	29
	152		14
	88		4

Figure 4.8. Mean and standard error of bottle oxygen data by depth bin for the whole Gulf (WG; green), Yucatan Channel (YC; red), and Bay of Campeche (BC; blue). The standard error of the mean is the black horizontal line at the mid-point of each bin. The number of data points, by bin, are given. The Bay of Campeche is south of and including 22°N latitude.

53

Table 4.4

Dissolved Oxygen Mean and Standard Deviation in mL·L^{-1} by Bin for the Whole Gulf of Mexico and the Yucatan Channel, Bay of Campeche, and Florida Straits Sub-regions

Region:	Whole Gulf			Yucatan Channel			Bay of Campeche			Florida Straits		
Bin Range (m)	NPTS	Mean	Std. Dev.	NPTS	Mean	Std. Dev.	NPTS	Mean	St. Dev.	NPTS	Mean	St. Dev.
0 – 50	7311	4.70	0.31	664	4.56	0.22	955	4.53	0.32	432	4.57	0.25
50 – 100	5670	4.40	0.59	479	4.42	0.33	602	4.31	0.60	249	4.29	0.46
100 - 150	3971	3.60	0.63	353	4.06	0.44	325	3.21	0.59	176	3.70	0.60
150 – 200	2797	3.26	0.43	322	3.66	0.26	257	2.80	0.24	152	3.36	0.43
200 – 250	2066	3.09	0.39	189	3.56	0.24	264	2.68	0.20	64	3.22	0.36
250 – 300	1732	3.03	0.44	204	3.60	0.24	196	2.54	0.16	70	3.21	0.30
300 – 350	1248	2.94	0.43	123	3.51	0.34	116	2.49	0.16	48	3.17	0.37
350 - 400	1151	2.88	0.38	119	3.31	0.28	147	2.50	0.14	50	2.98	0.25
400 – 450	1058	2.82	0.29	109	3.15	0.26	79	2.53	0.16	44	2.93	0.18
450 – 500	1050	2.85	0.26	125	3.08	0.23	128	2.59	0.14	52	2.82	0.24
500 – 550	727	2.85	0.24	114	3.00	0.21	78	2.67	0.17	37	2.91	0.20
550 – 600	597	2.92	0.20	80	2.95	0.17	92	2.81	0.18	25	2.95	0.21
600 – 650	544	2.95	0.24	67	2.94	0.20	35	2.89	0.21	16	2.94	0.33
650 – 700	470	3.12	0.28	72	2.96	0.24	48	3.05	0.24	17	3.07	0.35
700 – 750	515	3.20	0.27	62	3.04	0.20	70	3.20	0.26	15	3.23	0.27
750 – 800	620	3.35	0.31	74	3.17	0.29	88	3.39	0.21	24	3.21	0.36
800 – 850	429	3.49	0.35	53	3.33	0.29	20	3.58	0.28	15	3.23	0.27
850 – 900	199	3.64	0.36	51	3.48	0.33	27	3.76	0.27	8	3.55	0.48
900 – 950	273	3.78	0.31	48	3.67	0.33	43	3.81	0.27	9	3.84	0.32
950 – 1000	543	4.01	0.30	45	3.99	0.29	77	3.98	0.25	12	3.82	0.39
1000 – 1250	1159	4.35	0.36	134	4.36	0.37	124	4.34	0.32	22	4.43	0.30
1250 – 1500	608	4.77	0.25	83	4.82	0.29	105	4.69	0.26	13	4.73	0.30
1500 – 1750	345	4.89	0.24	66	5.03	0.26	18	4.87	0.22	9	4.91	0.29
1750 – 2000	233	4.96	0.31	49	5.16	0.32	24	4.75	0.32			
2000 - 2500	236	5.06	0.31	24	5.57	0.17	29	4.95	0.28			
2500 – 3000	152	5.03	0.20				14	5.04	0.24			
3000 - bottom	88	5.00	0.25				4	4.71	0.39			

Whole Gulf (n=35792)   NPTS = number of points
Yucatan Channel (n=3709): NW corner 22°N 87°W; SE corner 21°N 84°W (see Figure 4.7)
Bay of Campeche (n=3965): West of 90°W and south of 22.5°N exclusive
Florida Straits (n=1559): NW corner 25°N 83°W; SE corner 23°N 80°W (see Figure 4.7)

The Yucatan Channel profile also has oxygen concentrations at depths greater than about 1500 m that are high compared to the mean profiles within the Gulf. This is due to stations in the database that are located south of the sill. Data below about 2000 m in the Yucatan Channel are from stations located south of the sill. Waters at these depths are higher in oxygen than those at comparable depths within the Gulf because their source is the North Atlantic Deep Water with its high oxygen maximum core in the Caribbean of 5.5-6 mL·L^{-1} (Wüst 1964).

Second, there is a double oxygen minimum seen in the Yucatan Channel profile that indicates the presence of the oxygen maximum in the 18°C Sargasso Sea water near 300 m depth. Although this water mass may be present in individual profiles within the southeastern and, rarely, northeastern Gulf, it is not present in the mean in other profiles and even is weak in the mean profile for the

Florida Straits data within the Gulf (see also Morrison et al. 1983; Morrison and Nowlin 1977). This suggests 18°C Sargasso Sea water is eroded quickly as it mixes into the Gulf.

Third, the oxygen minimum of the Yucatan Channel is at approximately 600 m, which is deeper in than the whole Gulf at approximately 400 m or the Bay of Campeche at about 300 m. This is evidence of the general rise of isopycnals into the western Gulf. The depletion of oxygen in the whole Gulf and Bay of Campeche minimums relative to that of the Yucatan Channel is evidence of the age of the water. As the water mass circulates into the Gulf interior, the oxygen in this upper part of the water column is being consumed as organic matter decays. So, these profiles together suggest that the replenishment of the oxygen from the source waters into the Gulf interior is slow. The rates of replenishment, however, cannot be determined from the data available.

Examination of profiles from sub-regions of the Gulf of Mexico provides additional insights into the vertical structure of dissolved oxygen. For this purpose, the Gulf was divided into four quadrants, going counterclockwise: southeast, northeast, northwest, and southwest. Figure 4.9 shows the vertical profiles of the means for these quadrants and gives the standard error of the mean for each bin. Table 4.5 gives the mean, standard deviation, and number of data points.

The upper 1000 m exhibits the greatest differences between the four sub-regions. The southeast profile is similar to that of the Yucatan Channel, with the oxygen maximum associated with the 18°C Sargasso Sea water weakly evident at about 300 m and the deeper oxygen minimum at about 600 m. The mean profiles for the three other regions do not have the 18°C water maximum.

Moving counterclockwise through the quadrants from the southeast to the southwest, the oxygen minimum becomes shallower and the concentration at the minimum becomes less. This is evidence for the differences between the sub-regions in the replenishment rate of oxygen from the Loop Current system relative to the consumption rate in the Gulf interior. The mean profile for the southeast sub-region, much of which is influenced directly by the Loop Current approximately 70% or more of the time (SAIC 1989), is very similar to that for the Yucatan Channel (compare the Yucatan Channel column of Table 4.4 with the Southeast column of Table 4.5). For example, the oxygen minimum is 2.94 mL·L^{-1} at 600-650 in the Yucatan Channel and 2.96 mL·L^{-1} at 550-650 in the southeast Gulf. This strong match in the mean profiles indicates the southeast sub-region is replenished at a rate consistent with the rate of flows of the source waters into the Gulf.

The northeast sub-region, some of which also is influenced directly by the Loop Current but at less than 70% of the time (SAIC 1989), has less oxygen in the oxygen minimum (2.77 mL·L^{-1}), which itself occurs higher in the water column at 400-450 m. The mean profile in the northwest sub-region is similar in structure to that in the northeast sub-region, but with lower oxygen concentrations for given depths. For example, the oxygen minimum is at 400-450 m, as in the northeast, but has a lower mean of 2.65 mL·L^{-1}. This similarity likely reflects the movement of Loop Current eddies, which carry source waters, into the western Gulf. One analysis shows the presence of these eddies in the northwestern Gulf 5-15% of the time (SAIC 1989). The southwest quadrant has the lowest oxygen concentrations of all with the structure also higher in the water column (2.52 mL·L^{-1} at 350-400 m for the oxygen minimum). This decrease in oxygen concentration in the oxygen minimum is evidence that the oxygen concentrations from the source waters of the upper water column are being reduced by the local processes of oxygen consumption during organic decay as they move from the source region at the Yucatan Channel into the western Gulf.

To explore this, the western Gulf was divided into the Bay of Campeche (south of and including 22°N) and north-central western Gulf. Figure 4.10 shows profiles of means for the upper 1000 m of the two sub-regions. The standard error also is shown. The dissolved oxygen distribution in the Bay has an average minimum value significantly less than that in the north-central western Gulf (2.49 mL·L^{-1} at 300-350 m in Campeche Bay versus 2.63 mL·L^{-1} at 400-450 m in the north-central

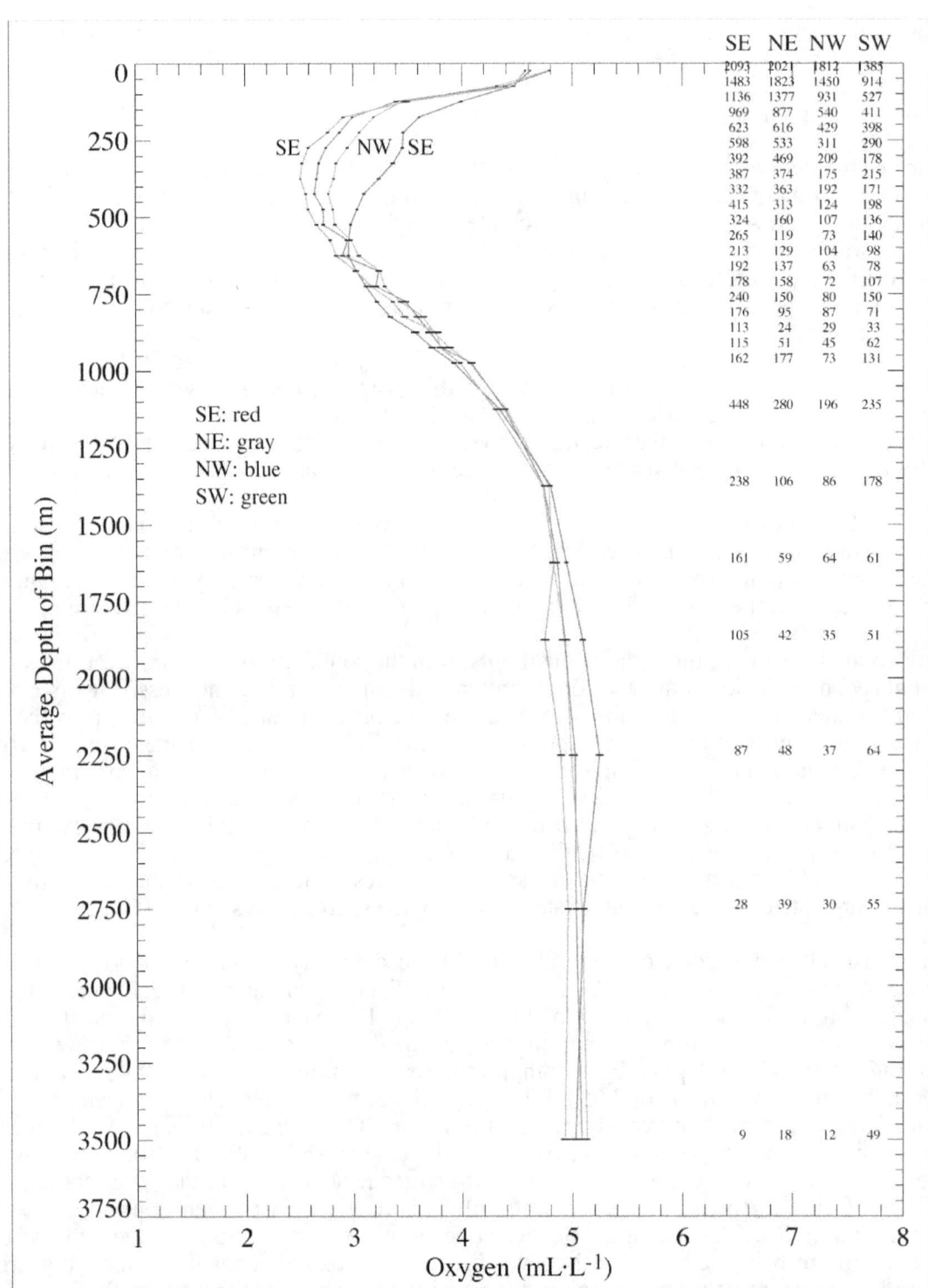

Figure 4.9. Mean and standard error of bottle oxygen data by depth bin for the southeast (SE: red), northeast (NE; gray), northwest (NW; blue), and southwest (SW; green) quadrants of the Gulf of Mexico. The standard error of the mean is the black horizontal line at the mid-point of each bin. The number of data points, by bin, are given. The boundaries of the quadrants are illustrated in Figure 4.7.

56

Table 4.5

Dissolved Oxygen Mean and Standard Deviation in mL·L^{-1} by Bin for the Southeast, Northeast, Northwest, and Southwest Sub-regions of the Gulf of Mexico

Region:	Southeast			Northeast			Northwest			Southwest		
Bin Range (m)	NPTS	Mean	Std. Dev.	NPTS	Mean	Std. Dev.	NPTS	Mean	St. Dev.	NPTS	Mean	St. Dev.
0 – 50	2093	4.62	0.26	2021	4.78	0.31	1812	4.80	0.27	1385	4.57	0.32
50 – 100	1483	4.46	0.43	1823	4.32	0.62	1450	4.39	0.69	914	4.46	0.58
100 - 150	1136	3.99	0.51	1377	3.46	0.51	931	3.40	0.65	527	3.50	0.75
150 – 200	969	3.61	0.37	877	3.19	0.32	540	2.99	0.28	411	2.91	0.31
200 – 250	623	3.47	0.33	616	3.06	0.28	429	2.88	0.22	398	2.77	0.26
250 – 300	598	3.46	0.35	533	2.95	0.29	311	2.75	0.18	290	2.59	0.18
300 – 350	392	3.37	0.40	469	2.84	0.28	209	2.69	0.16	178	2.54	0.18
350 - 400	387	3.24	0.34	374	2.82	0.24	175	2.67	0.15	215	2.52	0.14
400 – 450	332	3.10	0.29	363	2.77	0.18	192	2.65	0.15	171	2.57	0.16
450 – 500	415	3.04	0.24	313	2.82	0.17	124	2.73	0.17	198	2.59	0.15
500 – 550	324	2.98	0.22	160	2.83	0.20	107	2.73	0.17	136	2.67	0.18
550 – 600	265	2.96	0.18	119	2.98	0.19	73	2.95	0.22	140	2.79	0.18
600 – 650	213	2.96	0.24	129	3.05	0.24	104	2.90	0.24	98	2.85	0.20
650 – 700	192	3.03	0.29	137	3.25	0.25	63	3.24	0.22	78	3.03	0.26
700 – 750	178	3.12	0.28	158	3.30	0.22	72	3.22	0.26	107	3.17	0.27
750 – 800	240	3.23	0.34	150	3.44	0.30	80	3.49	0.27	150	3.37	0.22
800 – 850	176	3.35	0.33	95	3.65	0.35	87	3.59	0.33	71	3.48	0.29
850 – 900	113	3.57	0.39	24	3.74	0.33	29	3.71	0.25	33	3.76	0.27
900 – 950	115	3.73	0.34	51	3.88	0.30	45	3.81	0.26	62	3.80	0.25
950 – 1000	162	3.93	0.34	177	4.07	0.26	73	4.09	0.31	131	3.98	0.26
1000 – 1250	448	4.34	0.38	280	4.40	0.32	196	4.39	0.37	235	4.30	0.33
1250 – 1500	238	4.80	0.25	106	4.78	0.26	86	4.73	0.23	178	4.73	0.25
1500 – 1750	161	4.94	0.25	59	4.82	0.24	64	4.86	0.19	61	4.83	0.21
1750 – 2000	105	5.09	0.29	42	4.93	0.17	35	4.92	0.31	51	4.74	0.30
2000 - 2500	87	5.24	0.34	48	4.99	0.16	37	5.02	0.14	64	4.89	0.28
2500 – 3000	28	5.10	0.18	39	5.08	0.11	30	5.03	0.17	55	4.96	0.25
3000 - bottom	9	5.03	0.25	18	5.13	0.12	12	5.08	0.20	49	4.93	0.27

Southeast Gulf of Mexico (n=11482): East of 90°W inclusive and south of 25°N inclusive   NPTS=number of points
Northeast Gulf of Mexico (n=10558): East of 90°W inclusive and north of 25°N exclusive
Northwest Gulf of Mexico (n=7366): West of 90°W exclusive and north of 25°N exclusive
Southwest Gulf of Mexico (n=6386): West of 90°W exclusive and south of 25°N inclusive

western Gulf). Standard deviations are ~0.15 mL·L^{-1} and the standard errors are ~0.01 mL·L^{-1} suggesting that the means are statistically different. This is evidence of a minimal or slow exchange of waters from the northwestern Gulf with those of Campeche Bay, such as might occur in the presence of a persistent cyclonic feature in the Bay (Vázquez de la Cerda et al. 2005).

Mean profiles below 1000 m suggest oxygen levels are less at a specified depth the farther from the inflowing source waters a station is located. At the deepest bin (≥3000 m), however, there is a hint that the waters in the southeast sub-region may be lower in oxygen than those in the northeast or northwest. Unfortunately, this may be an artifact of the very small number of samples. If real, it likely would be related to the poorly understood deep circulation underlying the Loop Current (Nowlin et al. 2001).

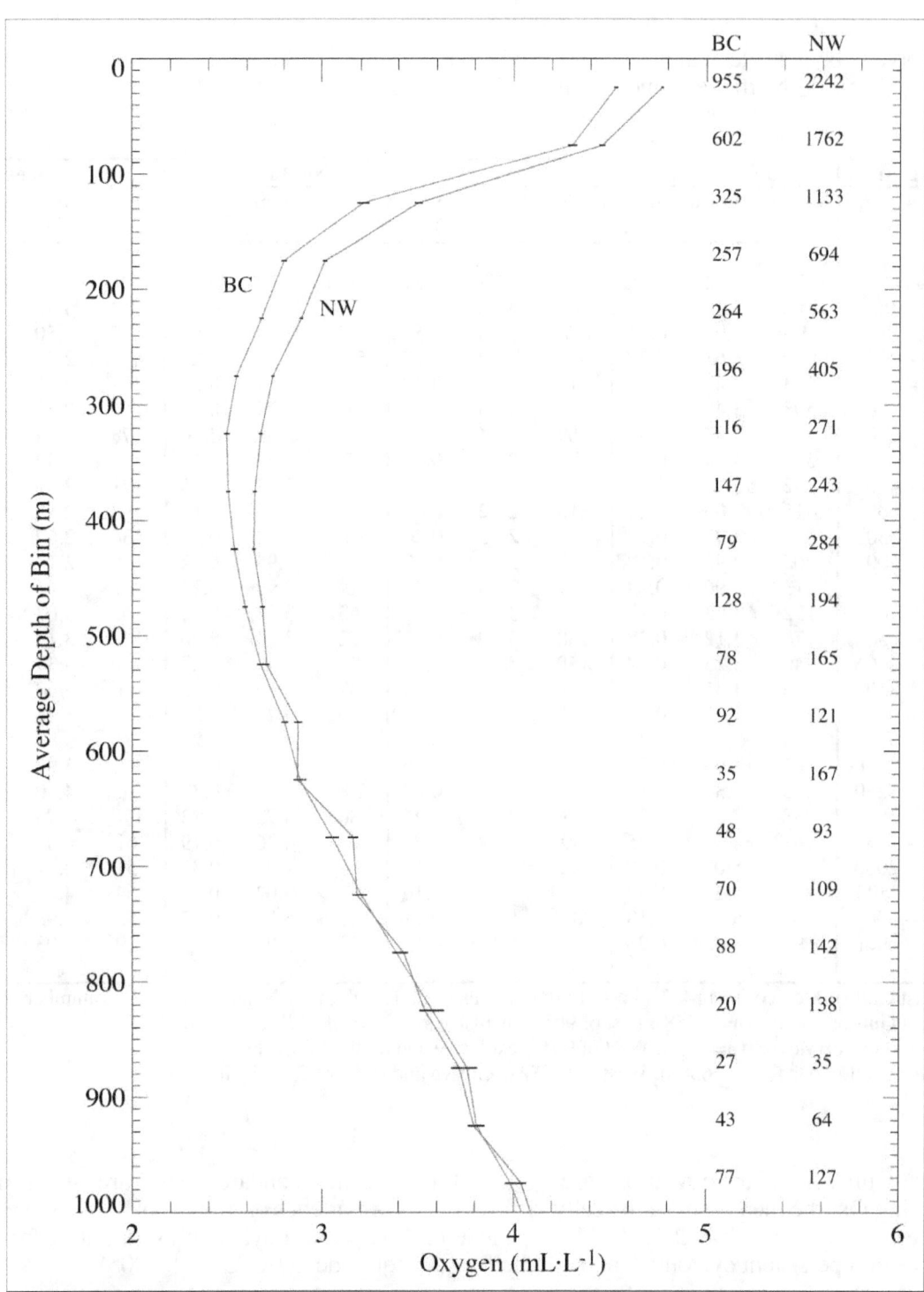

Figure 4.10. Mean oxygen concentrations by bins for the Bay of Campeche (BC; blue) and the northwest Gulf of Mexico (NW; red). All data used were west of 90°W, with the Bay of Campeche using those data south of and including 22.5°N and the northwest Gulf using all other data. The standard error of the mean is the horizontal line at the midpoint of each bin depth. The number of data points, by bin, are given.

Uniformity of Deep Concentrations: Nowlin et al. (1969) undertook a careful and detailed examination of dissolved oxygen data sets from the deep waters of the Gulf of Mexico. They were motivated by findings from three cruises in 1966 and 1967 that seemingly contradicted the study of McLellan and Nowlin (1963). That earlier study, based on data from the 1962 cruise 62-H-3 of the *Hidalgo*, had suggested there were large gradients in dissolved oxygen below approximately 1500-m depth. Nowlin et al. (1969) conducted a careful comparison of dissolved oxygen data from 62-H-3 and the 1964 cruises, 64-A-2 and 64-A-3, of the *Alaminos* with those from the 1935 cruise of the *Atlantis*, the 1958 and 1959 cruises of *Hidalgo* (58-H-4, 58-H-1, 59-H-2), and the 1966 and 1967 cruises of the *Alaminos* (66-A-8, 67-A-4, 67-A-8). They concluded that the data from the 1962 and 1964 cruises were based on faulty sampling or poor analysis techniques. They determined the data sets from the 1935, 1958, 1959, 1966, and 1967 cruises were good. Results of their analysis of these good data showed there was no clearly discernable horizontal variation in dissolved oxygen in the deep waters of the Gulf. Nowlin et al. (1969) argued this was consistent with the horizontal uniformity of salinity and potential temperature in the data from the 1966 and 1967 cruises and that McLellan and Nowlin (1963) found in the 62-H-3 data. They also found that there were only slight vertical gradients below the depth of the Yucatan sill, and that the mean dissolved oxygen concentrations of the deep waters at or below 1500 m was 4.99 mL·L^{-1} (excluding the 1962 data). Table 4.6 lists the mean, standard deviation, and standard error of the mean for each of the cruises examined by Nowlin et al. (1969). They found no evidence that oxygen concentrations had changed during the time period covered.

Table 4.6

Historical Statistics for Dissolved Oxygen At or Below 1500 m Depth in the Gulf of Mexico (after Nowlin et al. 1969; Wüst data are in the core of the NADW, based on data from 1932-1937)

Data set identifier	Mean (mL·L^{-1})	Standard deviation (mL·L^{-1})	Standard error (mL·L^{-1})	Number of data points
Atlantis 1935	4.96	0.134	0.028	23
58-H-1	5.08	0.124	0.028	19
58-H-4	4.87	0.054	0.020	7
59-H-2	5.01	0.055	0.013	17
66-A-8	4.99	0.048	0.011	19
67-A-4 Gulf basin	5.03	0.077	0.010	55
67-A-8	5.01	0.111	0.024	21
67-A-4 Yucatan basin	5.61	0.099	0.030	11
Wüst (1964) Yucatan basin	5.64	0.33	0.076	19
62-H-3	4.62	0.280	0.026	114
64-A-2 and 3	4.58	0.482	0.059	67

Statistics for data at or below 1500 m depth were computed using 1054 data points from 461 different stations that passed QA/QC processing. Figure 4.11 shows the locations of the stations. The statistics for selected sub-regions are presented in Table 4.7.

59

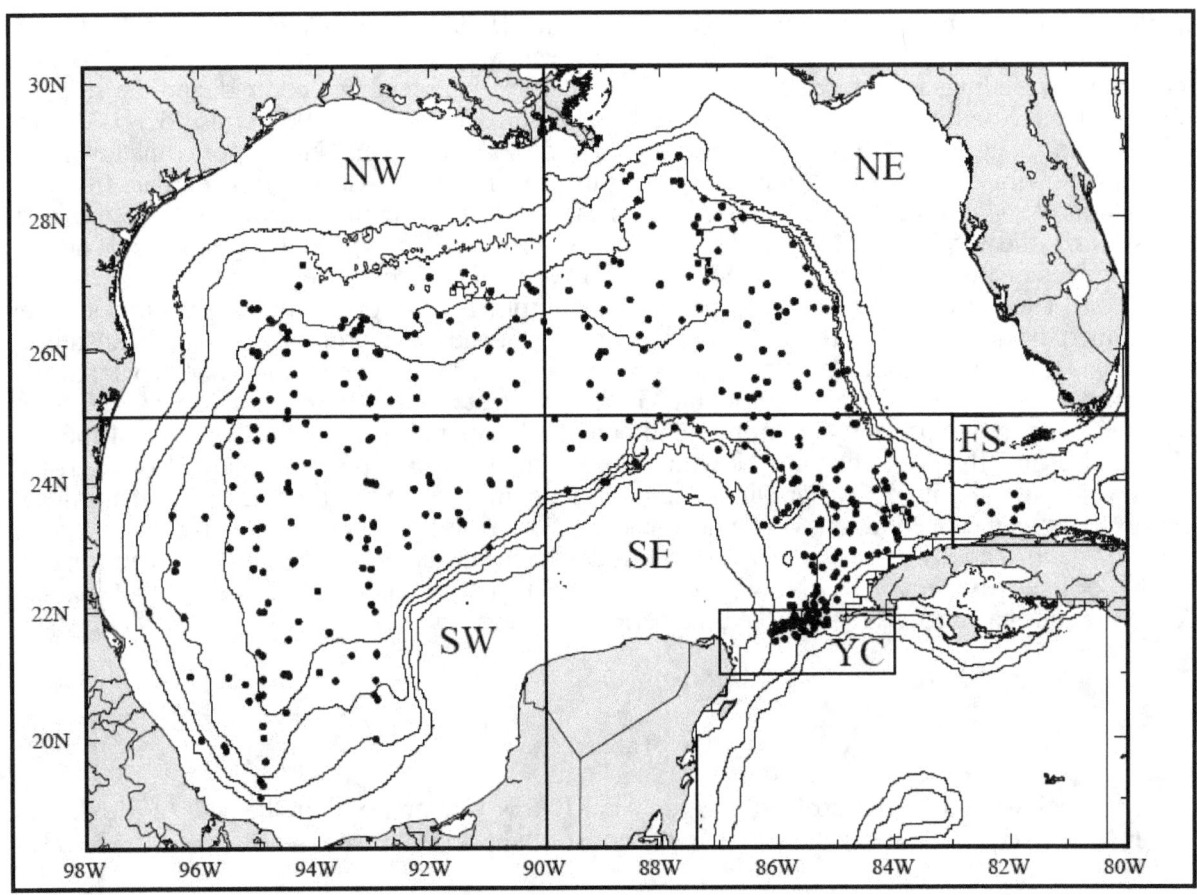

Figure 4.11. Locations of 461 stations with at least one good oxygen datum in depths ≥1500 m in the Gulf of Mexico. Geographical regions are shown, including the Yucatan Channel (YC) and the Florida Straits (FS). Bathymetry contours are 200, 1000, 2000 and 3000 m.

The mean dissolved oxygen concentration for the whole Gulf is 4.97 mL·L⁻¹, which is similar to the 4.99 mL·L⁻¹ found by Nowlin et al. (1969) for "good" data. The standard deviation, however, is approximately twice that of the Nowlin et al. (1969) study. This is due to the more rigorous data processing used in that earlier study as compared to this study, which rejected only the most clearly suspect data points. However, a rigorous processing was applied to the data base to provide a good data set for use in the box model (see Section 3.3). The mean of this data set is 4.98 mL·L⁻¹ with a standard deviation comparable to that of Nowlin et al. (1969). This and the standard errors of the means suggest that, although the standard deviations for the selected sub-regions are approximately 0.2-0.3 mL·L⁻¹, inferences based on the means can be reasonably drawn.

To examine the issue of horizontal uniformity of oxygen concentration below 1500 m, a first division compared the eastern with the western Gulf. The mean concentration in the eastern Gulf is greater than that in the western Gulf by 0.12 mL·L⁻¹. A Student-t test was applied to examine whether there was a statistically meaningful difference between the means in these two regions. The null hypothesis was

$$H_0: \mu_1 - \mu_2 = 0$$

where μ_1 and μ_2 represent the two population means. At the 95% confidence level, the null hypothesis is rejected, suggesting that the two regions have different mean dissolved oxygen concentrations at depth.

Table 4.7

Statistics for Dissolved Oxygen At or Below 1500 m Depth in the Gulf of Mexico from Data in the Oxygen Archive
(Regions of the Gulf are shown in Figures 4.7 and 4.11)

Region	Mean (mL·L^{-1})	Standard deviation (mL·L^{-1})	Standard error (mL·L^{-1})	Number of data points	Number of unique stations
Whole Gulf	4.97	0.27	0.008	1054	461
Whole Gulf – Box Model	4.98	0.09	0.009	97	43
East	5.03	0.28	0.011	596	268
West	4.90	0.26	0.012	458	193
Yucatan Channel above sill	5.09	0.30	0.028	116	61
Southeast outside YC	5.00	0.26	0.016	251	117
Northeast	4.96	0.21	0.015	206	90
Northwest	4.95	0.22	0.016	178	75
Southwest	4.87	0.27	0.016	280	118

The means of the deep water dissolved oxygen concentrations for the five sub-regions (Table 4.7) suggest the oxygen concentrations may decrease slightly going from the source waters in the Yucatan Channel into the southeast, northeast/northwest, and southwest Gulf. The greatest difference, of 0.22 mL·L^{-1}, is between the inflowing source waters at the Yucatan Channel and the waters of the southwest Gulf. Although this difference is comparable to the standard deviation, the standard error of the mean suggests these means are different. A Student-t test indicates that, at the 95% confidence level, the means are not from the same population. The means for waters of the northeast and northwest are within 0.01 mL·L^{-1} of each other, and the Student-t test supports that they are not different. Application of the Student-t test to the other pairs of sub-regions results in rejection of the null hypothesis of the same mean. The standard errors also support this, with the northeast and northwest values evidencing overlap in the means plus/minus the standard error and the southeast and southwest values evidencing no overlap. These statistics suggest that there may be a horizontal variation in the deep Gulf waters and that these waters may essentially be divided into three basic parts: southeastern Gulf, northern Gulf, and southwestern Gulf.

The relatively high mean concentration in the southeast Gulf is an indication that oxygen may be rapidly replenished from the source waters. The relatively low mean concentration in the southwest Gulf suggests these waters are not flushed as rapidly as the rest of the Gulf. However, from the standard deviations of 0.2-0.3 mL·L^{-1}, these statistics may also simply be the result of differences in oxygen titration methods, variations by analyst in the visual determination of the end point of the titrations, poor data collection, or poor sample analysis.

4.3.2   Patterns of Dissolved Oxygen Distribution

As seen in the statistical profiles, the fundamental distributions of dissolved oxygen are caused by the distribution of the inflowing source waters into the Gulf interior by the circulation with relatively small modifications caused by local inputs of oxygen in the photic zone and mixed layer and by oxygen consumption throughout the water column. These source waters consist of a number of water masses, including the oxygen minimum waters of the Tropical Atlantic Central Water (mean of ~2.9 mL·L^{-1} at 300-700 m depths) and the well-oxygenated waters of the Antarctic Intermediate Water (at 500-1000 m depths). Vertically, the structure everywhere within the Gulf consists of well-oxygenated waters (~4.5 mL·L^{-1} or higher) in the mixed layer that occurs approximately in the upper 100 m, declining concentrations to an oxygen minimum (generally < 3 mL·L^{-1}) ranging from ~300 m to 700 m, and increasing concentrations to well-oxygenated deep waters that are fairly uniform (~5 mL·L^{-1}) below about 1500 m. This pattern was seen by Morrison and Nowlin (1977) in data from the eastern Gulf on cruise 72-A-9 in May 1972 and by Morrison et al. (1983) in data from the western Gulf on cruise 78-G-3 in April 1978. Those papers provide detailed discussion of the vertical distributions of oxygen in the Gulf. Horizontally, dissolved oxygen concentrations in waters in the mixed layer are variable, depending as they do on the variable air-sea exchanges, photosynthetic activity, and proximity to regions with relatively heavy sedimentation (i.e., river outflows). Concentrations in waters below the mixed layer tend to decrease slightly the farther away the water is from the source waters inflowing at the Yucatan Channel, confirming that there is no water mass formation in the Gulf to replenish the dissolved oxygen consumed by the decay of organic matter. The source and sink system appears to have been in equilibrium over the 80 years of observations.

Structure of Dissolved Oxygen Distributions: Examination of vertical sections of dissolved oxygen concentrations with depth from selected cruises provides a means for visualization of the patterns of distributions seen in the mean profiles. To provide examples of the patterns of dissolved oxygen distributions, here are examined vertical sections from cruises in 1958 and 2000-2001.

In 1958, the research vessel *Hidalgo* conducted a number of cruises to examine the circulation and properties of the waters in the Gulf of Mexico. Results from two of these cruises, 58-H-1 and 58-H-4, are considered here. These cruises represent snapshots of the oxygen distribution associated with the then-current stage of the Loop Current intrusion into the eastern Gulf and Loop Current rings into the western Gulf. Figure 4.12 shows the locations of three transects from these cruises. Figure 4.13 is a vertical section of dissolved oxygen across the Yucatan Channel from cruise 53-H-4; all samples were collected on 18 May 1958. Figure 4.14 shows the distribution of dissolved oxygen concentrations in the eastern Gulf from the Mississippi River Delta to Havana, Cuba, collected 25-28 June 1958 on cruise 58-H-4. Figure 4.15 presents the vertical distribution in the western Gulf along –94.5°W longitude from data collected on 58-H-1 on 23-30 March 1958. These three transects are from McLellan (1960).

These three distributions have the same basic vertical distribution of dissolved oxygen throughout the water column. The dissolved oxygen values are relatively high near the surface, decrease to a minimum, and increase with depth below the minimum. Although the depth of the minimum is variable, being higher in the water column in the Yucatan Channel, the minimum occurs near the sigma-t of 27.1 on all three transects (see McLellan 1960). This exemplifies the tendency of water masses to flow along isopycnal surfaces, as well as the general uplift of those isopycnals from east to west in the Gulf.

The vertical section of dissolved oxygen for the Yucatan Channel (Figure 4.13) shows marked tilting of the isolines of dissolved oxygen in the upper 500 m. This tilting upward to the west of isolines also is present in the vertical sections (not shown here) of temperature, salinity, and sigma-t. It is indicative of the strong northward flow of the Yucatan Current in the upper waters that is

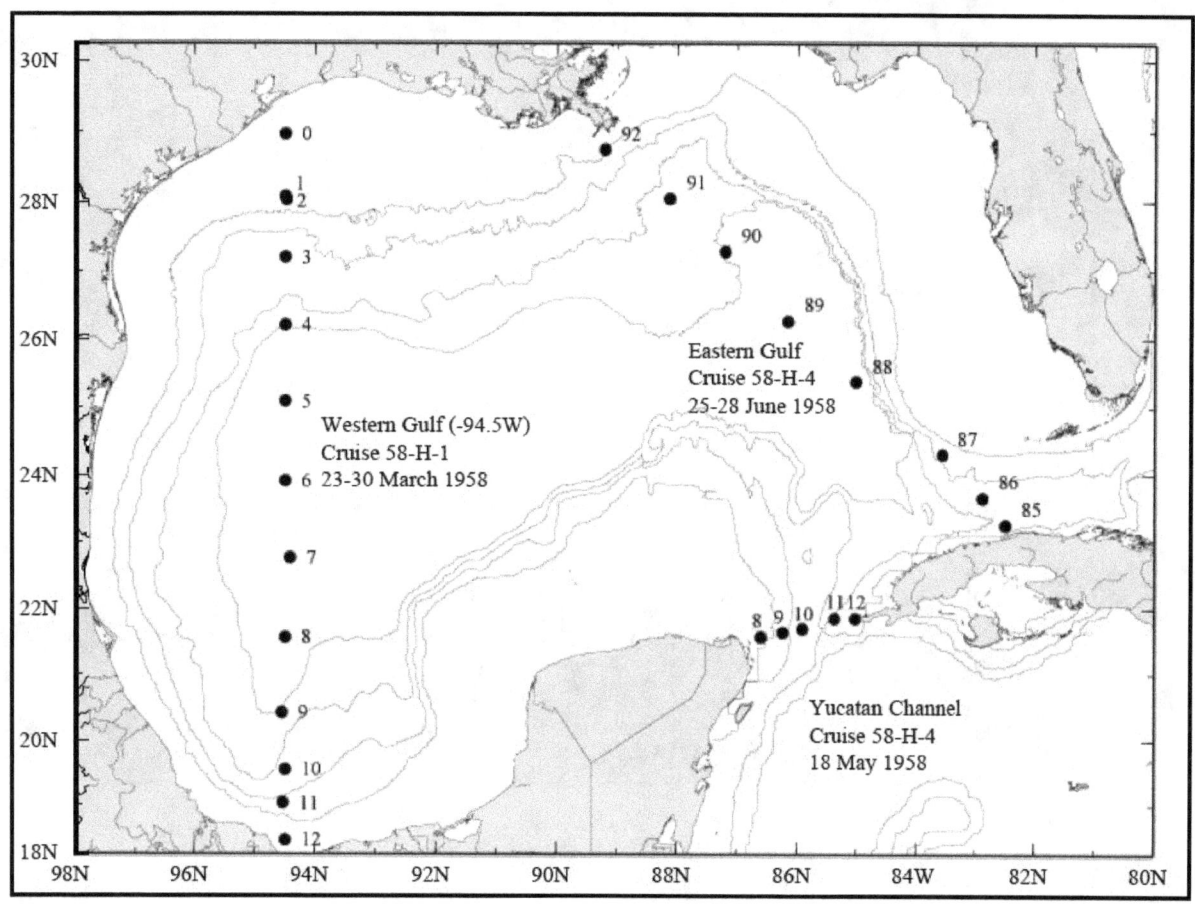

Figure 4.12. Station locations for Hidalgo cruises 58-H-1 and 58-H-4. Bathymetry contours are 200, 1000, 2000 and 3000 m.

intensified to the west. Note also that the sill depth of about 2000 m allows the well-oxygenated waters of > 4 mL·L^{-1} that are below about 1000 m depth to flow into the Gulf. These results are seen in other transects across the Yucatan Channel, including most recently that shown in Ochoa et al. (2001) from a data set collected in 1997. That recent transect also exhibits a pattern in the vertical distribution of dissolved oxygen that is similar to that in Figure 4.13, including the tilted isolines of oxygen and the presence of an oxygen minimum of < 3 mL·L^{-1} at depths of 300 m to 800 m.

The vertical distribution of dissolved oxygen in the eastern Gulf reflects the basic structure of the oxygen content of the source waters inflowing at the Yucatan Channel. Figure 4.14 also reflects patterns in distribution caused by a different circulation regime than that of the Yucatan Channel. There appears to be a clockwise circulation around station 90, as indicated by the depressed isolines at that station. There also appears to be an eastward flow south of station 88, again indicated by the depression in isolines. This structure suggests that either the Loop Current "made a wide sweep and entered the Florida Strait from the northwest" (McLellan 1960) or that a Loop Current Eddy had detached with its centerline located near station 90 and the Loop Current had turned eastward south of station 88. The result of this circulation is that, in the upper 800 m to 1000 m, the isolines of dissolved oxygen are deeper in the water column near the center of the feature at station 90 and deepen south of station 88. The oxygen minimum had values generally greater than 2.7 mL·L^{-1} at

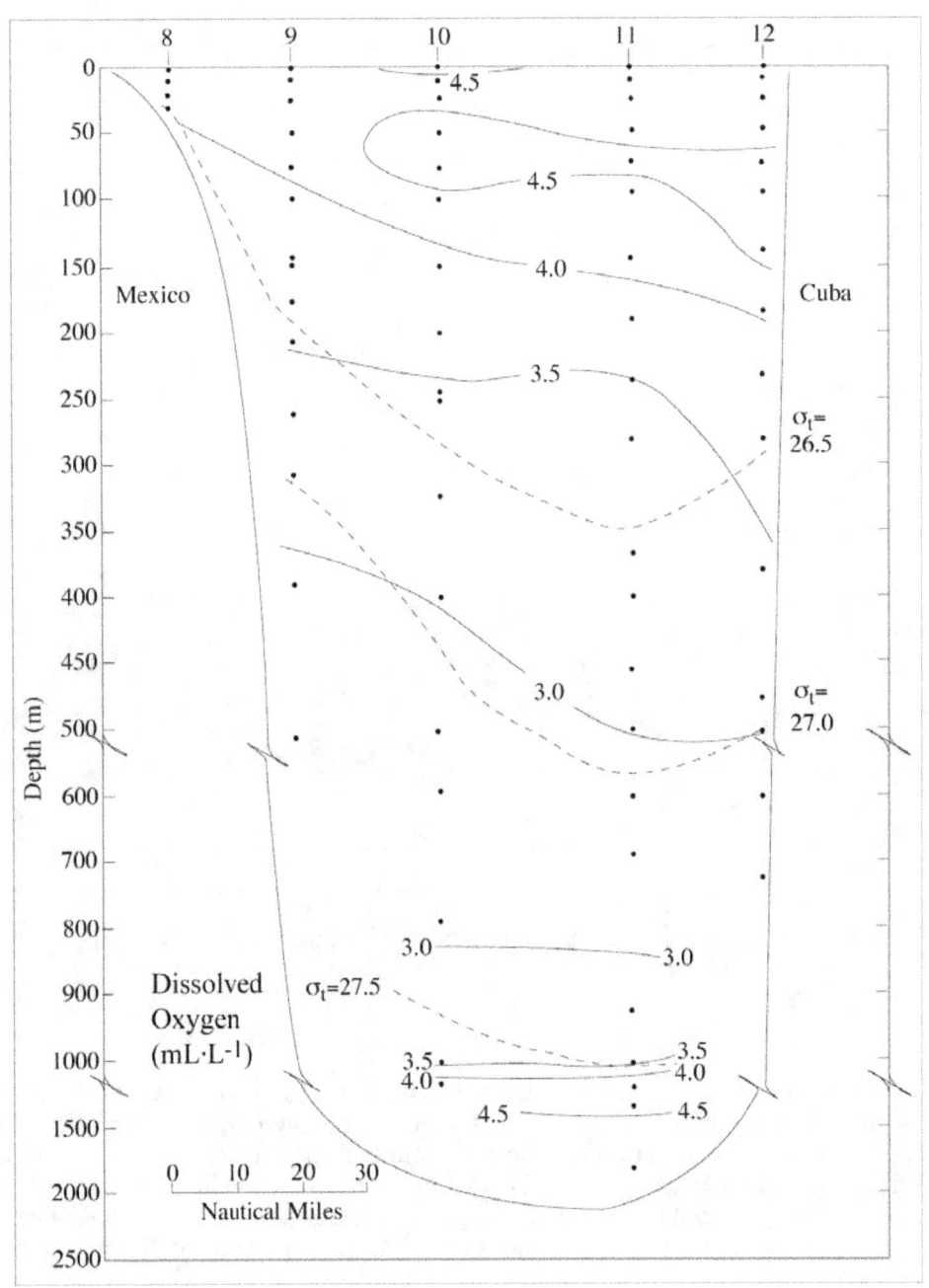

Figure 4.13. Vertical distribution of dissolved oxygen across the Yucatan Channel, Cruise 58-H-4, 18 May 1958. (after McLellan 1960)

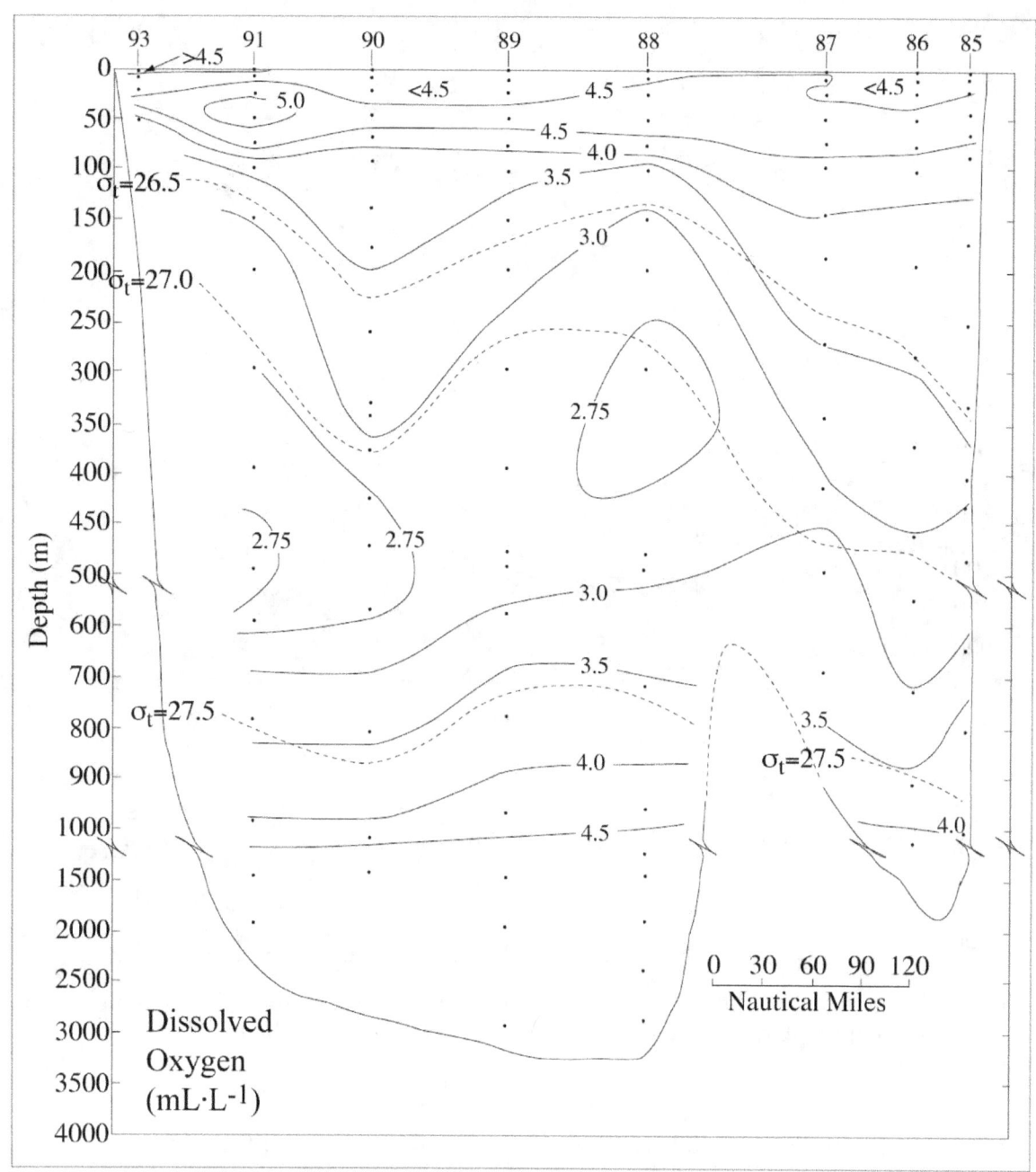

Figure 4.14. Vertical distribution of dissolved oxygen in the Eastern Gulf, from the Mississippi River Delta to Havana, Cruise 58-H-4, 25-28 June 1958. (after McLellan 1960)

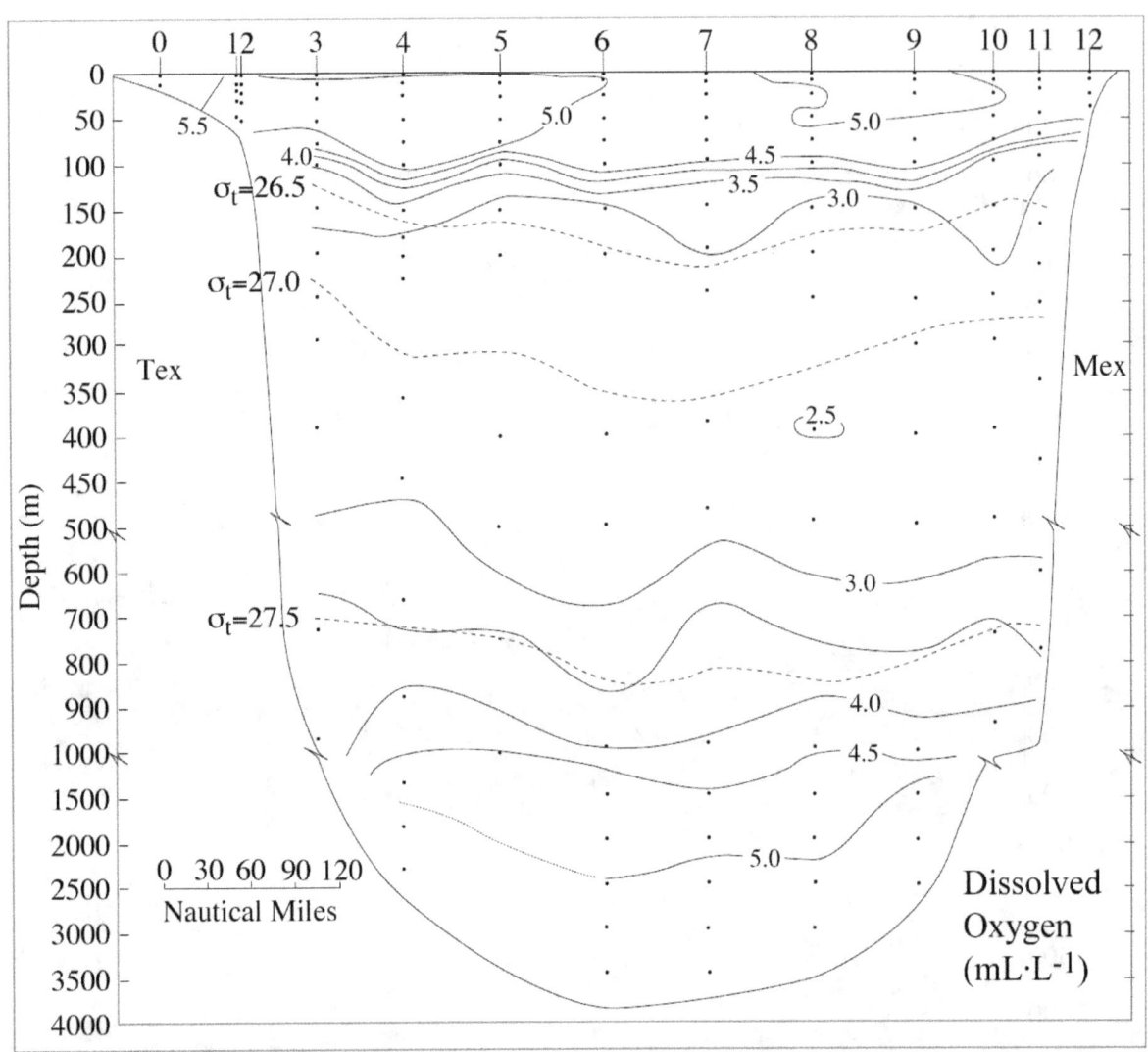

Figure 4.15. Vertical distribution of dissolved oxygen in the Western Gulf, along -94.5°W longitude, Cruise 58-H-1, 23-30 March 1958. (after McLellan 1960)

depths ranging from ~350 m to 500 m at different parts of the section due to the circulation. Below about 1000 m the waters are well-oxygenated, with the highest concentration being 4.9 mL·L^{-1} near bottom.

In contrast to the eastern Gulf in June 1958 (Figure 4.14), the western Gulf in 1958 had no evidence of strong circulation features such as Loop Current Eddies, that depress (anticyclones) or raise (cyclones) isolines. The temperature, salinity, and sigma-t sections (not shown) suggest a general anticyclonic circulation with an eastward counterflow near the Mexican coast and a westward flow on the Texas shelf (McLellan 1960). The isopleths of dissolved oxygen exhibit only small vertical uplifts and depressions, rather than the strong depressions or uplifts that were seen in association with the strong circulation features of the eastern Gulf. The oxygen minimum in the western Gulf had values generally from 2.5 to 2.8 mL·L^{-1} at sigma-t level of about 27.1 and depth of ~400 m. This pattern differs from that of the eastern Gulf; this reflects the differences in circulation, as well as the decrease in oxygen concentrations due to the greater net consumption of oxygen that occurs in the western Gulf because those waters are farther from the source waters than those of in the eastern Gulf. Maximum oxygen concentrations below 2500 m were > 5.0 mL·L^{-1}.

The near surface values of the three transects exhibit a number of differences, which likely reflect different processes at work. Seasonal differences in the air-sea exchanges are suggested by comparison of the relatively lower, more stratified oxygen values (~4.5 mL·L^{-1}) of the upper waters in the summertime transect in the eastern Gulf (Figure 4.14) with the relatively higher, well-mixed concentrations (~5 mL·L^{-1}) of the early spring transect in the western Gulf (Figure 4.15). Of the three transects, the maximum oxygen values near surface were > 5.9 mL·L^{-1} and occurred over the Texas shelf in the western Gulf transect (Figure 4.15). These waters had temperatures that were relatively cold (~16°C) and salinities that were relatively low (< 35 to as low as 32.7), exemplifying the typical wintertime conditions in this region that are favorable for increased oxygen uptake by the ocean in air-sea exchanges. Variability in photosynthetic activity is indicated by the local sub-surface oxygen maximum at about 50-100 m depth at station 91 offshore of the Mississippi River Delta (Figure 4.15). This would be an area that could have higher summertime primary productivity, with increased photosynthesis, associated with nutrient inputs from the Mississippi River (see e.g., Jochens et al. 2002).

The MMS-sponsored project, DGOMB, provided an opportunity to sample much of the northern Gulf to near bottom. Data from two cruises were examined: Cruise 1 in 4 May – 17 June 2000 and Cruise 2 in 2-18 June 2001. Figure 4.16 shows the sea surface height during each cruise with the stations from a transect selected from each cruise. The June 2000 cruise crossed the northern half of a Loop Current Eddy that was in the western Gulf (Figure 4.16 upper panel). The dissolved oxygen section for this transect (Figure 4.17) shows the characteristic deepening isolines from the outer edge (station RW1) to the center (station RW6). The basic vertical structure remains the same, as seen in the *Hidalgo* cruises and as found by Morrison and Nowlin (1977) for the eastern Gulf in 1972 and Morrison et al. (1983) for the western Gulf in 1978. The near-surface waters are well-oxygenated, as is typical for waters that have contact with the atmosphere and photosynthetic activity. The core of the oxygen minimum occurs on the sigma-theta surface of 27.15. The well-oxygenated deep waters occur below about 1000 m. Oxygen concentrations below 1500 m were ≥ 4.9 mL·L^{-1} for this transect. The June 2001 cruise sampled the eastern Gulf. Although there were a number of eddy features (anticyclonic and cyclonic) in the area sampled, no transect crossed such features (Figure 4.16 lower panel). A transect consisting of the selected stations from the cruise was constructed and is presented in Figure 4.18. The station spacing is too wide to allow conclusions about the effects of circulation on the dissolved oxygen distributions. However, this figure shows, once again, the consistent pattern of the vertical structure of dissolved oxygen in the Gulf. The oxygen minimum again is on the sigma-theta surface of 27.15. Note the high oxygen concentrations for the deepest waters, below the sill depth of the Yucatan Channel; these exceed 5 mL·L^{-1} at the samples below 1500 m. This is indicative of the continued replenishment of dissolved oxygen to the deep waters from the only source available–overflowing waters from the Yucatan Channel.

Patterns from oxygen/sigma-theta plots: The water masses tend to have their characteristic property cores along specific isopycnals (see, e.g., Wüst 1964; Wennekens 1959; Morrison and Nowlin 1977; and Morrison et al. 1983). The data sets from the May/June 2000 and June 2001 DGOMB cruises on the R/V *Gyre* were compared to those from the 1972 *Alaminos* cruise 72-A-9 examined by Morrison and Nowlin (1977) and the 1978 *Gyre* cruise 78-G-3 examined by Morrison et al. (1983). DGOMB data were split into east and west to match, respectively, the 72-A-9 and 78-G-3 data sets from those two regions. Figure 4.19 shows the dissolved oxygen concentrations versus sigma-theta, with the black symbols being from the older cruises and the red data being from the DGOMB cruises.

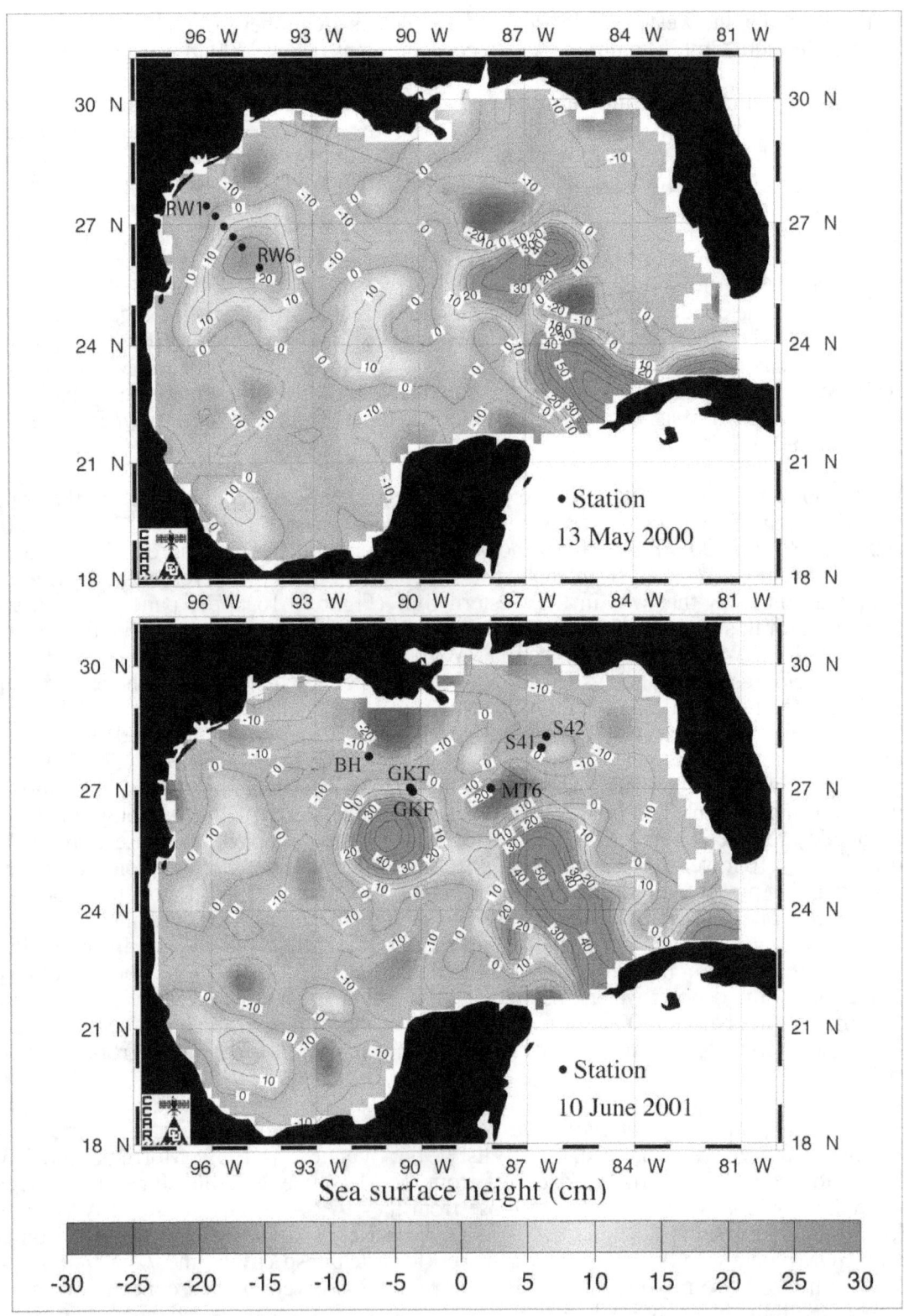

Figure 4.16. Sea surface height for TOPEX/ERS2 analysis for 13 May 2000 and 10 June 2001. Station locations for transects shown in Figures 4.17 and 4.18 are shown.

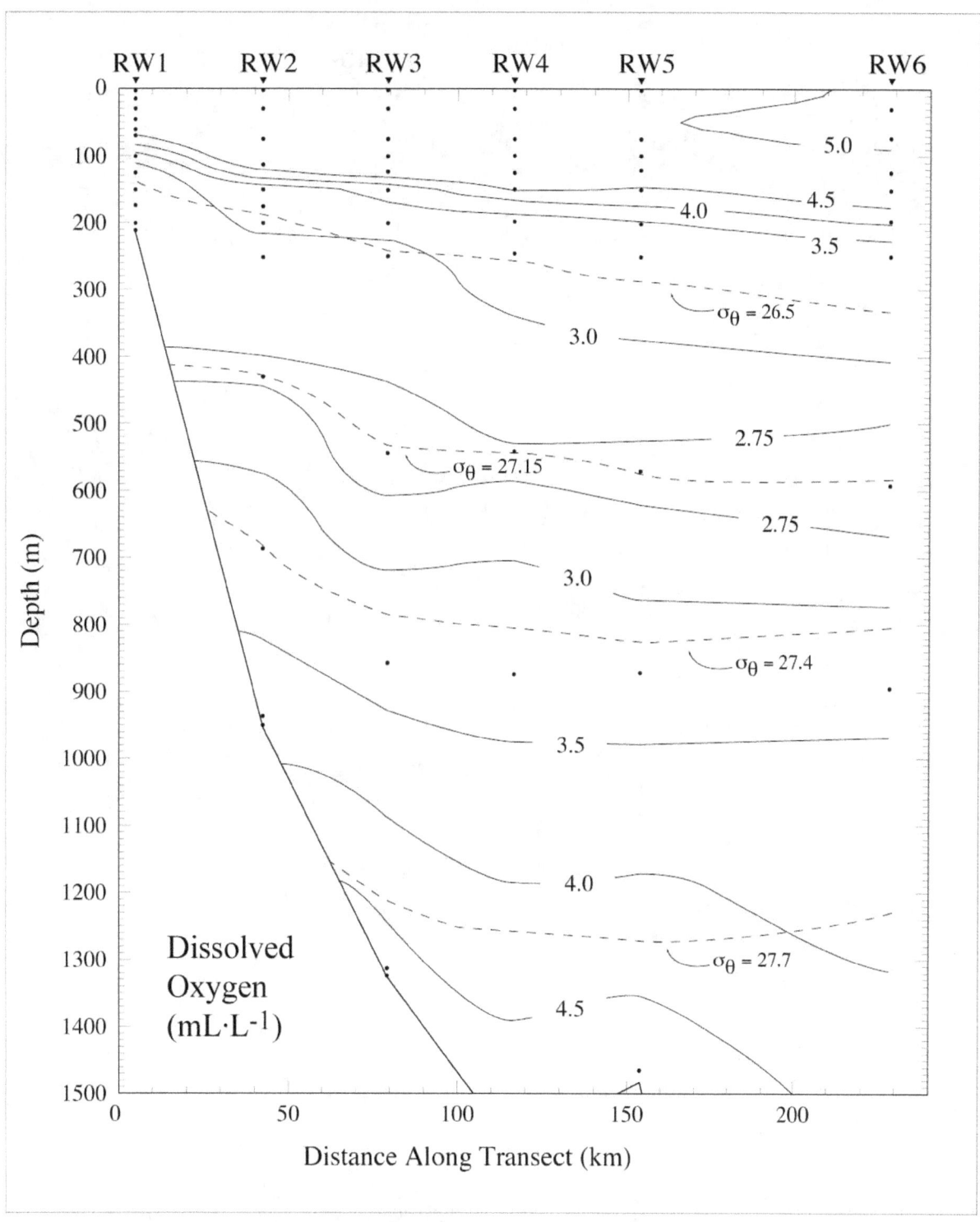

Figure 4.17. Dissolved oxygen on the western-most transect of DGOMB cruise DG1, 4 May - 17 June 2000. Dashed line shows the sigma-theta density surface associated with selected water masses (Table 4.1). Dots show sampling locations.

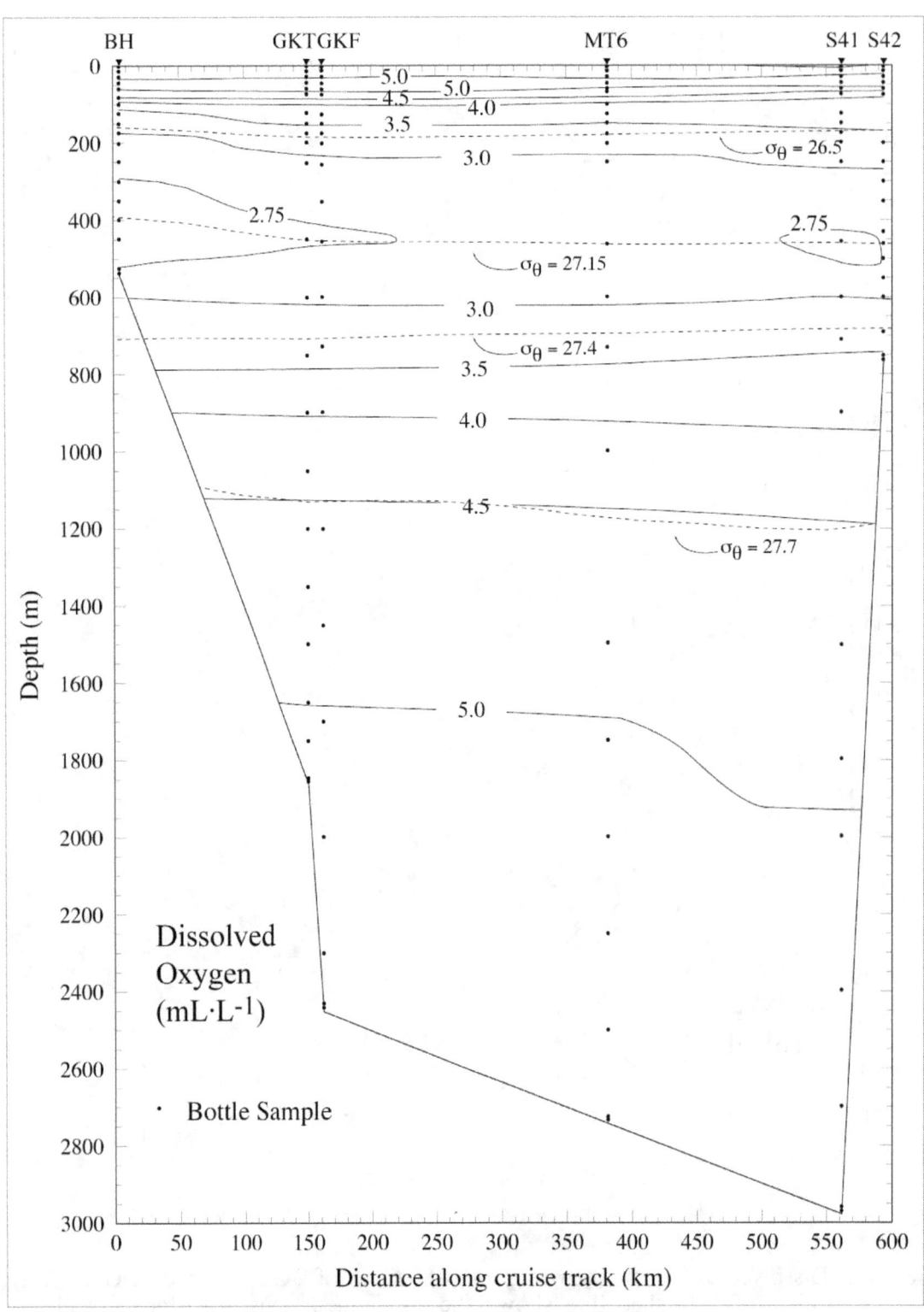

Figure 4.18. Dissolved oxygen on DGOMB cruise DG2, 2-18 June 2001. Dashed lines show the sigma-theta density surfaces associated with selected water masses in the Gulf (Table 4.1). Station locations are shown in Figure 4.16.

70

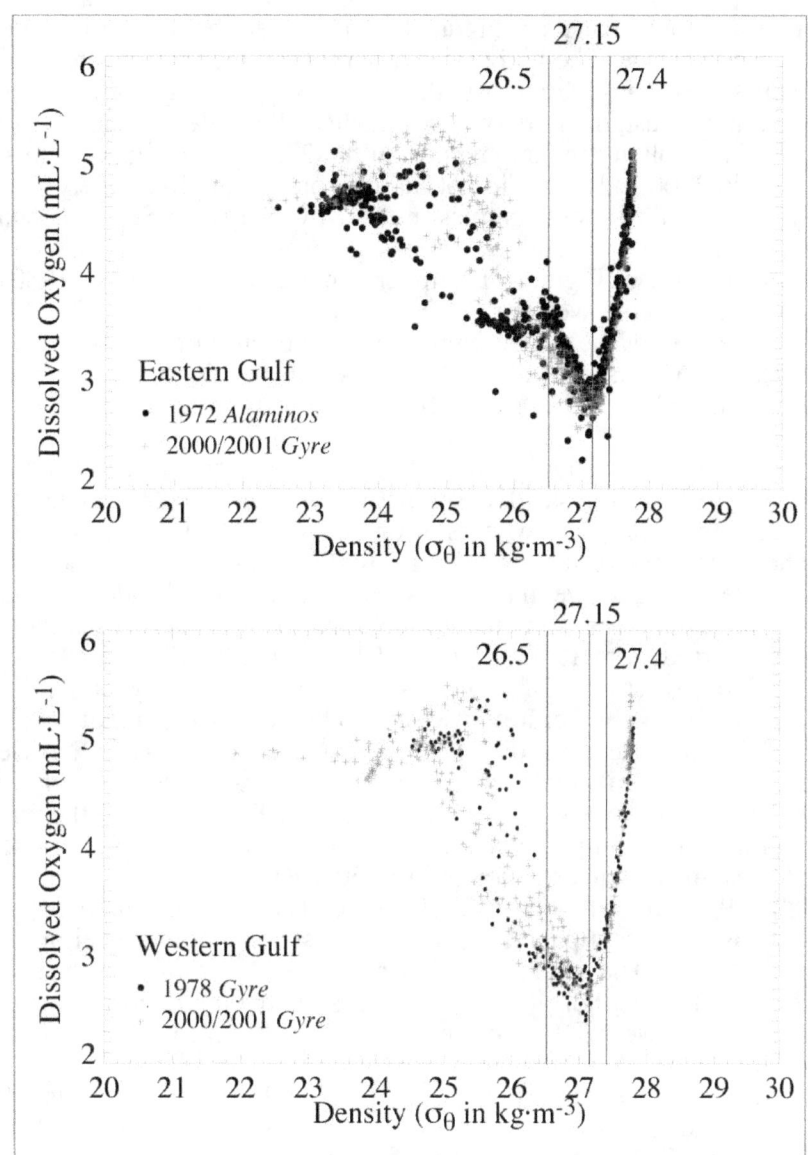

Figure 4.19. Dissolved oxygen on density surfaces for the eastern and western Gulf of Mexico. Data from the east Gulf are located at or east of -90°W; data from the west are west of -90°W. Specific cruises shown are the May 1972 Alaminos cruise (72-A-9), April 1978 Gyre cruise (78-G-3), and May/June 2000 and June 2001 Gyre cruises (00-G-05 and 01-G-05) of the DGOMB project. Shown are sigma-theta density surfaces for the 18°C Sargasso Sea Water (26.5 kg·m^{-3}), Tropical Atlantic Central Water (27.15 kg·m^{-3}), and Antarctic Intermediate Water (27.4 kg·m^{-3}).

Comparing the upper and lower panels of Figure 4.19, it is seen that the basic patterns between the east and west Gulf are the same. The horizontal variability in the upper waters is large, reflecting differences in the air-sea exchanges, photosynthetic inputs, and oxygen consumption rates of the near-surface waters. In contrast, the horizontal variability of the deep waters, below the Antarctic Intermediate Water with a salinity minimum core on the 27.4 kg·m^{-3} sigma-theta surface, is small. The DGOMB cruises in 2000 and 2001 did not sample south of approximately 27°N in the eastern Gulf. This likely resulted in the lack of samples in the 18°C Sargasso Sea Water that occurs along sigma-theta surface of 26.5 kg·m^{-3}. This water mass, however, is evident in some of the stations from the 1972 *Alaminos* cruise (Figure 4.19, upper panel). Morrison et al. (1983) found that this water mass does not occur in the western Gulf; this is confirmed by the DGOMB data (Figure 4.19, lower panel). The core of the oxygen minimum has been found by Wüst (1964), Morrison and Nowlin (1977), and Morrison et al. (1983) to be on the 27.15 kg·m^{-3} sigma-theta surface throughout the Gulf. The DGOMB data set confirms that this is still the case.

Temporal Variations: In their detailed comparison of oxygen data sets from 1935, 1958, 1959, 1962, 1964, 1966, and 1967 (see 4.3.1 above), Nowlin et al. (1969) found no evidence that dissolved oxygen concentrations from waters at or below 1500 m had changed during the time period covered. The comparison of the data sets from 1972 and 1978 with those from 2000-2001, as well as the results described above in the statistical section, provide additional evidence that no changes have occurred, even higher in the water column. To consider this issue further, the data were divided into six periods: pre-1950, 1950i to 1960, 1960i to 1970, 1970i to 1980, 1980i to 1990, and 1990i to 2001, where the "i" indicates that data from this year were included for the period. Data then were divided into east and west and binned as before, and then statistics were computed. Figure 4.20 shows the mean dissolved oxygen values by decade. No clearly discernible temporal trends are present. For example, considering the deepest values, in both the east and west, the decade of 1960-1970 is lowest in mean concentration, 1990-2001 is in the middle, and 1970-1980 is highest. For another example, considering the oxygen minimum bin of 450-500 m in the eastern Gulf, the lowest mean concentration is in 1950-1960 (with the smallest sample size), the next lowest is 1980-1990, and the highest is 1970-1980. The oxygen minimum bin of 300-350 m in the western Gulf has a different pattern, with the lowest concentration being in 1970-1980 and the highest being 1950-1960. These results together with results from other analyses in this study and from the early investigators indicate that the processes that maintain the oxygen distributions have not changed over the 80 years encompassed by the available data sets.

Influence from shelves on deep ocean oxygen distributions: To examine effects of shelf processes on the deepwater oxygen distributions, the northern Gulf was divided into the 2° longitude swaths of 86-88°W, 88-90°W, 90-92°W, 92-94°W, and 94-96°W. The data within these swaths north of 25°N then were separated by total water depth into regions bounded by the 200-1000 m, 1000-2000 m, and 2000-3000 m isobaths. Data in the upper 1000 m then were binned in 50-m deep bins and the mean and standard deviation for each bin were computed. The pattern of oxygen distribution in the 86-88°W swath (Figure 4.21), which includes DeSoto Canyon, suggests that lower oxygen concentrations might occur in the shallower total water depths at and above the oxygen minimum zone. The swath for 88-90°W shows a similar order, but with less difference between the relatively high and low oxygen regions. However, because the deep water regions in these swaths encompass territory that is directly influenced by the Loop Current, this result more likely is due to the occurrence of higher oxygen concentrations from those source waters in these deepwater regions (2000-3000 m and 3000 m-bottom) than from influence of the shelves. Additionally, this pattern is not clear when the standard deviations, which are large, are considered. Below the oxygen minimum in this swath and for all swaths west of 90°W, there is no discernible pattern (Figure 4.21). Thus, little if any influence from the shelves are seen in the oxygen distributions in the northern Gulf.

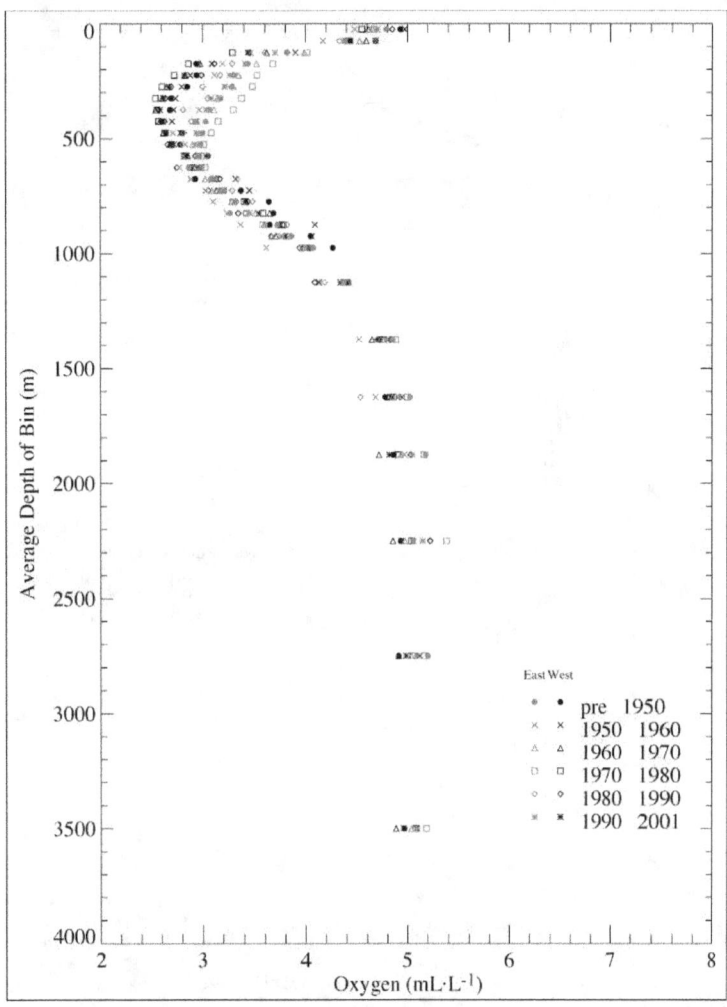

Figure 4.20. Mean dissolved oxygen in the Gulf of Mexico by decade. Data from the eastern Gulf (red) are located east of and including -90°W. Data from the western Gulf (black) are located west of -90°W.

73

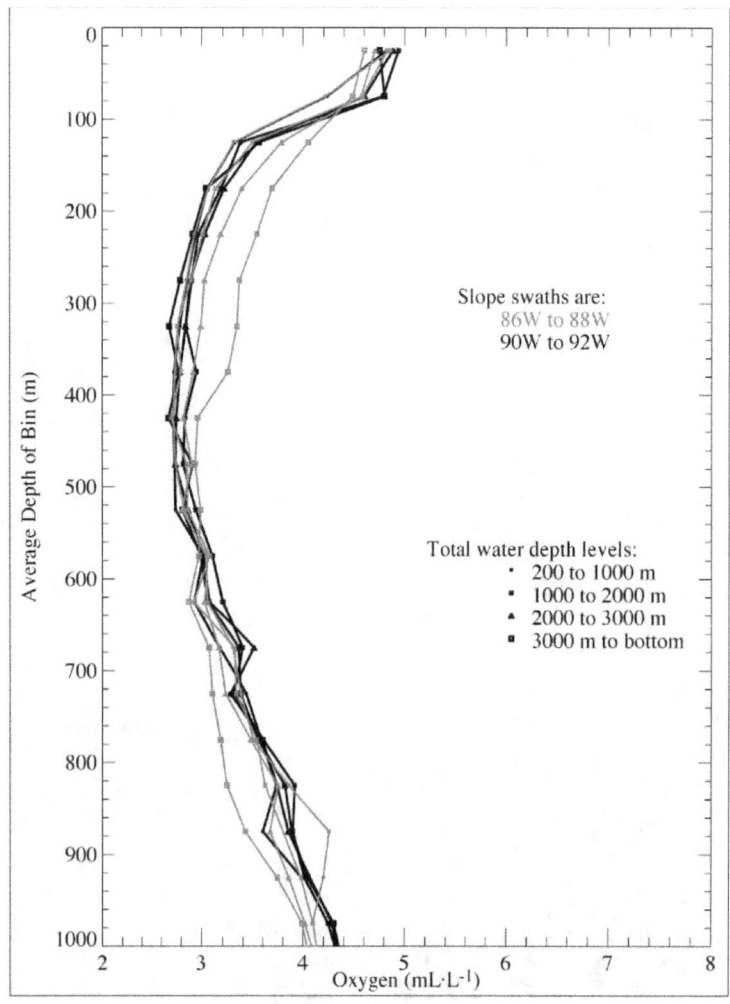

Figure 4.21. Mean dissolved oxygen in 50-m depth bins for selected total water depths in the northern Gulf of Mexico. Only data north of 25°N were used.

The LATEX A (Nowlin et al. 1998a, 1998b) and NEGOM (Jochens et al. 2002) programs provide information on the dissolved oxygen distributions over the shelves and insights into possible exchanges with the deepwater. As with Nowlin and Parker (1974), the LATEX A and NEGOM programs showed no water mass formation over the shelves. This is for the simple reasons that the shelf waters never become cold or saline enough to create the dense waters necessary to flow off the shelf and ventilate the deep waters. Profiles along the shelf edge (LATEX A and NEGOM) and 1000-m isobath (NEGOM) indicate that the upper water masses, such as the Subtropical Underwater and the Tropical Atlantic Central Water, may be present at the shelf edge and that circulation features such as cyclonic and anticyclonic eddies, squirts, and jets may result in exchange of waters between the shelf and the open ocean (see Sahl et al. 1997; Nowlin et al. 1998a, 1998b; Jochens et al. 2002). However, these exchanges are limited to the upper few hundred meters and do not extend into the deep waters.

The results of the LATEX A and NEGOM programs show that the processes over the shelves have little influence on the oxygen distributions in the upper waters of the deepwater Gulf. Any influence on the oxygen content of the lower waters would have to be from transport of organic material off

74

the shelf into the deep basins, but the oxygen data do not suggest that this contributes in any unusual way to the normal oxygen consumption patterns.

These results suggest there is no strong influence of shelf processes on dissolved oxygen concentrations in the deepwater Gulf. Rather, the effect of the processes at the continental margin on the oxygen distributions in the open ocean waters seems to be related to processes of off shelf circulation interactions with the topography, such as uplift of isopycnals from east to west in the upper 1000 m in response to the general circulation pattern of the Gulf.

4.4   Effects of Oil and Gas Inputs on Dissolved Oxygen Concentrations

A natural hydrocarbon spill experiment in the Gulf of Mexico has been ongoing for millions of years in the form of naturally occurring seeps that introduce oil at many locations throughout the basin. The mean dissolved oxygen concentrations in the deep Gulf have not noticeably changed over the last 80 years, and the stability of the oxygen distribution in the deep Gulf suggests the natural experiment has not resulted in large depletions in the oxygen reservoir of the deep Gulf (see Section 4.3). For this reason, the Gulf of Mexico system must be considered in equilibrium with regard to dissolved oxygen concentrations, with ventilation processes balancing local consumption processes. For example, on a basin scale, the Gulf dissolved oxygen system has not shown any effect from the ~150,000 tonnes of hydrocarbons typically input to the Gulf each year from natural and anthropogenic sources. However, on local spatial and short-term scales, the effects of such inputs could be profound.

In this section, we discuss how oil and gas inputs can affect the concentration of dissolved oxygen in the deepwater region of the Gulf of Mexico. Rather than an exhaustive listing of every process that can potentially affect dissolved oxygen concentration, we limit our discussion to those processes considered most significant in deep water. We begin the section with a discussion of the dissolved oxygen content of the Gulf of Mexico as a function of water depth. Next, is a general discussion of the sources of petroleum in the ocean followed by a discussion of the general processes that can affect dissolved oxygen concentration in the water column, including the effects of drilling fluids, drill cuttings, and discharge of produced water. A brief summary of the IXTOC I and other case study spills is next, followed by a set of simple calculations of the impact on dissolved oxygen concentrations by extreme oil input into the ocean. Last, a simple model is presented to quantify the amounts of oil needed to produce specified changes in the oxygen content of the Gulf of Mexico.

Background: A series of three reports by the National Research Council details our progressive understanding of the sources, fates, and effects of petroleum in the ocean (NRC 1975, 1985, 2003). Information from the 1985 and 2003 reports was used in considering the effects on dissolved oxygen concentration of oil and gas inputs, although this issue was not directly considered in these reports. These reports detail the many sources of petroleum in the ocean: anthropogenic inputs from extraction, transportation, and consumption of petroleum and natural hydrocarbon seeps. From available data for 1990-1999, the average annual inputs of petroleum hydrocarbons to the Gulf from anthropogenic sources are approximately 2000 tonnes from extraction with 1700 tonnes being from produced waters, 1600 tonnes from transportation, and 6800 tonnes from consumption (NRC 2003). These anthropogenic discharges can occur either at specific sites (e.g., extractive inputs), or over wide areas (e.g., consumption inputs). Most of these inputs are near surface and so have little likelihood of directly impacting the dissolved oxygen in the deep waters.

Natural oil seeps in the offshore Gulf of Mexico are estimated to contribute approximately 95% of the total oil inputs (NRC 2003). The seepage rates are estimated to be 140,000 ± 60,000 tonnes per year (NRC 2003), based on an estimate of Mitchell et al. (1999) of 70,000 ± 30,000 tonnes per year for the northern Gulf. However, these natural discharges occur in specific regions and usually have relatively low and chronic release rates.

The geologic record indicates that natural hydrocarbon seepage has been active in the Gulf of Mexico for tens of millions of years (MacDonald 2002). At the sites of this natural leakage of hydrocarbons, localized low oxygen conditions can occur in the upper meter or so of the sediments and, hence, likely would influence oxygen concentrations of the waters at the sediment-water interface. This effect is caused by oxygen consumption by the chemosynthetic communities that live off the hydrocarbon seeps (see MacDonald et al. 1995; MacDonald 2002). However, dissolved oxygen concentrations measured in the water column within a few cm above seeps are at levels typical of the bottom waters (MacDonald and Schroeder 1993). Additionally, concentrated brine pools are associated with the seeps, and these pools are anoxic (see MacDonald 2002). However, these are localized features that do not appear to have impacted the general oxygen concentrations in the deepest waters of the Gulf. Chemosynthetic communities are active at the seep sites. Such communities require access to oxygenated waters because chemosynthesis, which is a pathway for carbon fixation, consumes oxygen (MacDonald 1992; MacDonald et al. 1995; MacDonald 2002). In the Gulf, this access is provided by the well-oxygenated bottom waters (MacDonald et al. 1995). Thus, the chemosynthetic communities in the Gulf are additional evidence of ventilation of the deepest Gulf waters.

The relative inputs from the various sources, shown in Figure 4.22, indicate that typical, annual hydrocarbon inputs to the Gulf are dominated by the natural sources. As seen in Section 4.3, the dissolved oxygen concentration in the deep waters of the Gulf are maintained by a balance of oxygen inputs from circulation of well-oxygenated deep waters inflowing from the Caribbean Sea at the Yucatan sill and oxygen consumption. This balance is maintained even in the presence of a substantial natural input of hydrocarbons to the system. Although these inputs do not appear to have perturbed the dissolved oxygen system of the basin, they can have extreme (positive and negative) effects locally (see, e.g., discussions of chemosynthetic communities that live off the oil seeps in the Gulf; MacDonald 1992, 2002).

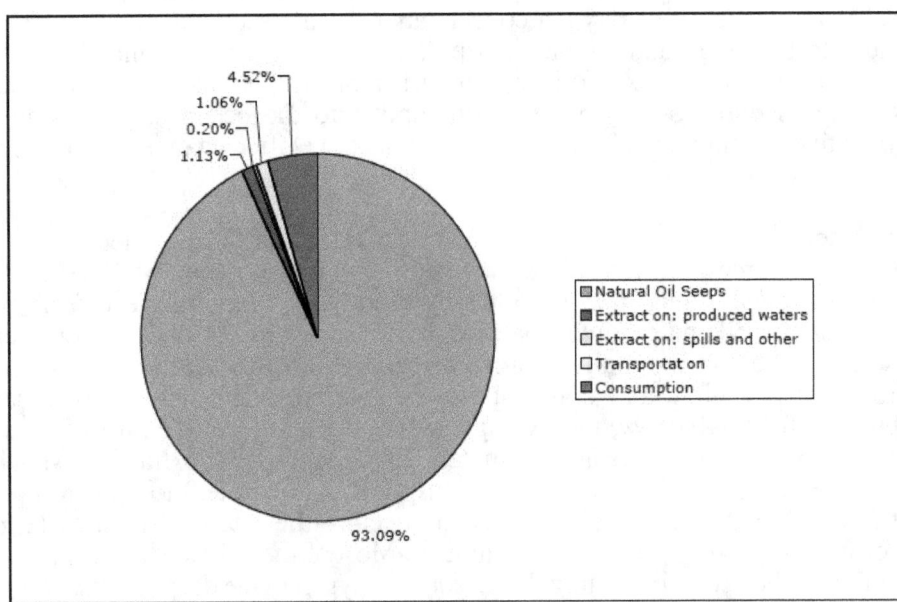

Figure 4.22. Percent of annual average hydrocarbon discharge by natural and anthropogenic sources. Approximate discharges are: 140,000 tonnes from natural oil seeps, 1,700 tonnes from extraction-produced waters, 300 tonnes from extraction-other, 1,600 tonnes from transportation, and 6,800 tonnes from consumption (NRC 2003).

76

The input of hydrocarbons from anthropogenic sources generally are small compared to the natural input and so also are not expected to perturb the dissolved oxygen system of the deep Gulf. Nevertheless, they can be expected to have possibly extreme local effects. Furthermore, catastrophic oil spills, such as IXTOC I (discussed below), can introduce quantities of hydrocarbons that are several times greater than the inputs from natural sources (Figure 4.23). As discussed below, the effects of such a massive spill on the total dissolved oxygen system of the Gulf does not cause a major perturbation to that system. However, local effects on dissolved oxygen concentrations can be extreme as the hydrocarbons released and the organic matter from the resulting mortality decay. The factors involved with these local processes, both from typical anthropogenic discharges and from catastrophic spills are many and complex.

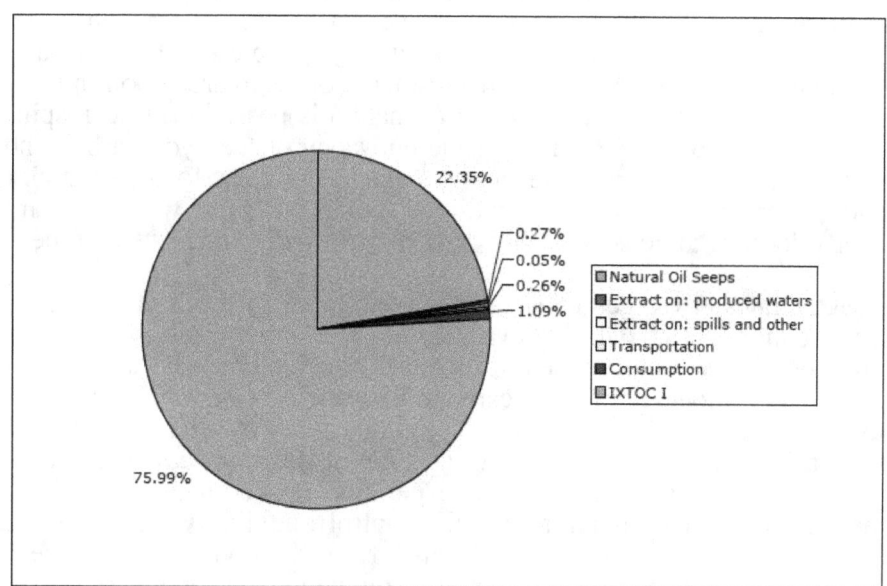

Figure 4.23. Percent of annual average hydrocarbon discharge by natural and anthropogenic sources compared to the discharge from the IXTOC I oil spill. Approximate discharges are: 476,000 tonnes from the IXTOC I spill, 140,000 tonnes from natural oil seeps, 1,700 tonnes from extraction-produced waters, 300 tonnes from extraction-other, 1,600 tonnes from transportation, and 6,800 tonnes from consumption (NRC 1985; NRC 2003).

Processes that affect the fate of hydrocarbon discharges into the oceanic environment include weathering (evaporation, emulsification, and dissolution), oxidation (chemical and biological), and transport (advection and spreading, dispersion and entrainment, sinking and sedimentation, partitioning and bioavailability, and shoreline stranding with possible tarball formation). Each of these processes is examined in detail by the NRC (2003).

The NRC (1975, 1985, 2003) reports also provide detailed information on the effects of hydrocarbons on the oceanic environment, primarily to biological marine communities or individual organisms. NRC (2003) notes that the level of impact depends on a complex set of factors, such as rate of release, type of petroleum, physical properties at the site of discharge and/or contact with land, and local biological characteristics of the ecosystems that are exposed. Of particular interest to our study is the effect on dissolved oxygen concentrations in the water column of hydrocarbons discharged into the deepwater oceanic environment of the Gulf of Mexico. It is expected that the rates of oxygen consumption due to hydrocarbon oxidation would vary for each of the factors described above and that the environment would vary for different parts of the water column in a deepwater release. To assess the effects on dissolved oxygen would require a complex oil spill model beyond the scope of this study.

Another important factor in assessing effects is whether the oil spill is at or near the sea surface or is a subsurface release. The vertical structure of subsurface discharges extends from the point of discharge to the sea surface. It is thought to consist of a jet phase near the discharge point at the sea floor, then a plume phase with its own complex dynamics in the next several hundred meters, followed by a post-terminal phase where the rise is driven by the buoyancy of individual drops or particles and associated hydrodynamic drag, and eventually a surface slick phase where the hydrocarbon reaches the surface (see NRC 2003 for details). NRC (2003) reports that most of the oil from a deep discharge would rise to the surface over several hours although some oil would be dissolved and/or emulsified; most if not all of the natural gas, if present, likely would dissolve into the water. This would impact the longevity of the residence of the hydrocarbon in the water column, and so its availability for oxidation. Also hydrate formation is possible for deep spills with natural gas present; formation of hydrates would lower the buoyancy of the hydrocarbon and increase the amount of time the hydrocarbon takes to reach the surface. However, Johansen et al. (2002) report that hydrates did not form as predicted in a controlled spill in 844 m of water 125 km off the coast of Norway. Clearly, local conditions again will determine the nature and extent of the impacts.

It is expected that the rates of oxygen consumption would vary for each of the different phases of flow in the water column, but no data are available on what these rates would be in each phase. However, decay takes time to occur, so the impact in the water column on the oxygen concentration will depend highly on how long labile hydrocarbons are present in the water column. It is likely that most of the hydrocarbon will rise to the surface within hours of release, so that direct oxygen consumption most likely will occur at the surface. Emulsification may retain a portion of the hydrocarbon in the water column for extended periods, but, coupled with the transports by circulation, the chemistry of such mixtures will control and likely slow the rate of oxygen consumption within a particular region. The distance from the point of discharge to the surface determines the extent of the region in which mixing of oil with water can occur and how much time is available for the soluble oil fractions to dissolve into the water column (NRC 2003). Once at the surface, different processes, including clean-up, act on the hydrocarbon. At the surface, the air-sea exchange of oxygen, which is impacted by the hydrocarbon slick, can offset part or all of the oxidation processes that remove oxygen from the water.

Effect on dissolved oxygen concentrations in the water column: The input of oil and gas can affect dissolved oxygen concentration in the water through photo-oxidation as well as microbial oxidation (NRC 2003). The photo-oxidation process is limited by the presence of sunlight and the optical qualities of the incident light, water, and hydrocarbons. It is necessarily a near-surface process. The importance of this process for consumption of oxygen at the surface is unclear. However, NRC (2003) reports it is unimportant in a mass balance sense. It will have no effect in the deep waters.

A more important process impacting oxygen concentrations is microbial oxidation. It acts throughout the water column, and so directly affects oxygen concentrations in the deep waters. The extent to which microbial oxidation occurs is dependent on many environmental factors including water temperature, chemical composition of the oil, abundance and species of microorganisms, available dispersal mechanisms for hydrocarbons that might remove them prior to their oxidation, available nutrients (nitrogen and phosphorus), salinity, and pH (NRC 1985; Von Oudenhoven et al. 1983). Oxidation rates of hydrocarbons are mostly limited by their solubility and dispersion.

Oil can be dispersed in the ocean by several mechanisms. As a result, for oil spills in the open ocean, most of the oil does not come ashore, but rather dissipates *in situ*. In the open ocean, 10-75% of spilled oil can be evaporated from the surface within a few days (NRC 2003). This again, however, is dependent on many parameters such as chemical composition of the petroleum, meteorological and oceanographic conditions, and the surface thickness and water depth of the spill. Higher wind speeds and air and water temperatures greatly increase the rate at which oil evaporates (Chapman 1985). Temperature can affect both the rate of evaporation at the surface and the rate of microbial growth and activity and therefore, the biodegradation rate which in turn affects the oxidation rate at all levels within the water column. Differences in biodegradation rates exist even among different types of hydrocarbons. For example, tars and resins are typically resistant to biological attack, while paraffins, alkanes, alkynes, and monoaromatics are readily biodegradable (e.g., Von Oudenhoven et al. 1983; NRC 1985).

At the surface, the interaction of waves, winds, and currents can lead to the formation of patches of a thick foamy substance called mousse. This consists of an emulsion of oil in water, with the water component typically being about 80-90%. The formation of the mousse is believed to slow the biodegradation processes of the oil. Because of this, the formation of mousse is considered to be a mechanism for transport of hydrocarbons over long distances. It is thought that the formation of mousse may also slow the degradation process by limiting the amount of light that can penetrate the surface layer and by slowing evaporation rates. Gibbs (1976) notes that mousse also limits biodegradation by reducing the rate of diffusion of oxygen and nutrients into the interior of the mousse volume. In addition, mousse produces much thicker surface layers. Oxygen limitation of biodegradation of oil slicks at the surface is thought unlikely because there is direct contact with the atmosphere. However, oxidation limitation may occur when the oil is dispersed in the water column, but nutrient limitation of biodegradation is probably more likely (NRC 1985).

The oxygen concentration can be further lowered as biological organisms associated with the oxidation of the hydrocarbons die and decompose in the water column and in the benthos. Again, this is unlikely to be important in well-oxygenated water, but could be important if the oil sinks, particularly in stratified water close to shore (i.e., shallow waters) where organic sediments exist and the oxygen concentrations are low.

Because of the wide range of hydrocarbon types and differing environmental conditions, the rate of oxygen consumption by oxidation is not clear. However, a rate of oxidation of oil in seawater was found by ZoBell (1969) to be 0.2 to 2 mg oil per liter per day at 20°C and at concentrations of 0.5 to 50 ml oil per liter of seawater. Adding oil above 50 ml per liter did not change the rate of oxidation. This compares to oxidation rates reported by Gibbs (1976) of 0.082 mg oil per liter per day at 14°C (0.03 mg oil per liter per day at 4°C). The latter values assume a theoretical oxygen demand of 3.5 g O_2 per g of oil. It is also well known that the rate of biodegradation increases as the size of dispersed oil droplets gets smaller, presumably because of the increased surface area for bacterial settlement and activity. The microbes must access the dissolved oxygen and nitrogen from the water as well as the carbon in the oil (P. Chapman, personnel communication).

Effect of drilling fluids on dissolved oxygen concentrations: Drilling fluids can affect local concentrations of dissolved oxygen through organic enrichment near the seabed. There are three major categories of drilling fluids: water-based muds, oil-based muds, and synthetic-based muds. Locally, the impact of oil-based drilling fluids on benthic communities can be severe and toxic (see e.g., Avanti Corporation 1997a); discharge of oil-based muds and associated cuttings has been prohibited in the U.S. waters. Discharge of water-based muds and associated cuttings is allowed. Although synthetic-based drilling muds themselves are not allowed to be discharged, the drill cuttings may be rinsed and then discharged into the ocean with a residue of the mud remaining on them. Impacts on dissolved oxygen concentrations from drilling fluids have two sources: impacts from the biodegradation of any organic material in muds or associated cuttings that may be

79

discharged and physical accumulation of cuttings (or muds if discharged) at the seafloor that smothers benthic communities which then degrade. The organic material content of the muds and cuttings, and their solubility and bioavailability, and hence the possible effects on dissolved oxygen concentrations, varies with the characteristics of the mud used and the formations drilled.

The impact of drilling muds and cuttings on dissolved oxygen concentrations in the water column and at the seafloor also will be highly dependent on the local environmental conditions, including the water depth, energetics of local currents and waves, substrate type, and location of the discharge point (e.g., near-surface or near sea floor), as well as the nature and volume of the discharges (Avanti Corporation 1997b). Neff et al. (2000) reviewed drilling fluids, with emphasis on synthetic-based muds, and their environmental risks. They found that cuttings piles rarely occurred in deepwater or high-energy marine environments, but that where muds and cuttings accumulated (field studies examined were in water depths <600 m), adverse effects occurred to the local benthic communities. The drilling muds and cuttings smother the benthic communities that are down-current of the discharge point. The resulting organic material undergoes aerobic degradation processes that lead to oxygen depletion in the water column immediately above the sediments and depletion to hypoxic or anoxic conditions in the surficial sediment layers (Avanti Corporation 1997a; Neff et al. 2000). Finally, there can be an emergence of sediment anaerobic microbial communities (Avanti Corporation 1997a; Neff et al. 2000). Synthetic-based drilling-muds are believed to have high oxygen demands during biodegradation, but there is as yet no consensus on their biodegradation (Avanti Corporation 1997a). Neff et al. (2000) report that, relative to synthetic-based fluids, biological effects are likely to be limited to sediments near the discharging platform, with the organic enrichment and resulting oxygen depletion as the main mechanisms of adverse impact, and that within 3-5 years of cessation of discharge the oxygen concentrations will have increased to levels that would allow recovery of the ecosystem affected. Rates of oxygen consumption are unclear and highly dependent on localized factors.

Summary of the IXTOC I case: When considering the effects of an oil spill on the environment, it is instructive to consider one of the most famous and most studied of all oil spills in an ocean environment. The IXTOC I production platform of PEMEX was on the continental shelf of Campeche Bay in the southern Gulf of Mexico. From 3 June 1979 to 23 March 1980, approximately 476,000 tonnes of crude oil were spilled from this platform into the open ocean over the continental shelf, making the IXTOC I spill the second largest marine spill event (NRC 2003), second only to the deliberate release of oil from multiple sources during the 1991 Gulf War. The water depth at this location was 52 m; the spill occurred 14 m above bottom. The oil rose to the surface forming a plume that emanated from the spill site. At the spill site a fire cone 25-30 m in diameter consumed some of the released oil. The slick was estimated at 1 mm thick and 300 m wide and stretched several hundred miles northwest to the Texas coast, following wind driven currents (NRC 1985). Oil concentrations in the water column during September 1979 ranged from 10.6 $mg \cdot L^{-1}$ a few hundred meters from the blowout to less than 0.005 $mg \cdot L^{-1}$ at a distance of 80 km from the blowout (Boehm and Fiest 1982). Comparatively, maximum oil concentrations from the Ekofisk blowout in the North Sea and *Amoco Cadiz* and *Argo Merchant* tanker spills were 0.30, 0.35, and 0.45 $mg \cdot L^{-1}$, respectively. These values were comparable to concentrations found at the edge of the oil plume of IXTOC. The higher concentrations at IXTOC are attributed to the subsurface release (Boehm and Fiest,1982).

Approximately 10% of the spilled oil from IXTOC I was recovered. Microorganisms consumed 30 to 50% of the spill, and the balance of the oil (40-60%) was removed through weathering processes such as dispersal in the water column or evaporation to the atmosphere. Between 0.5 to 3% of the spilled hydrocarbons remained in the bottom sediments near the spill. About 3000 to 4000 metric tons of oil washed onto the beaches of southern Texas. At the time of the spill there was no information on how the local biota were affected by the spill because there were no pre-spill data with which to compare post-spill data. It was observed that within 25 km of the spill, local microbe communities were replaced with hydrocarbon utilizing populations. However, nutrients were limiting, so little degradation was noted (NRC 1985).

Worst case computations of impact: ZoBell (1969) found that the aqueous dissolved oxygen in 320,000-400,000 L of seawater would be consumed by just one liter of oil, i.e., in English units one barrel of oil would oxidize 40 acre-feet of seawater. For perspective, the IXTOC I spill was 3,332,000 barrels, while the *Exxon Valdez* oil spill was 262,000 barrels. ZoBell's value, based purely on molecular balancing of chemical equations, makes many assumptions that are practically unrealistic. The assumptions include complete mixing of the oil with the entire volume of the available seawater, unlimited time for the oxidation to take place, absolutely no contact with the atmosphere for oxygen replenishment, and no processes such as burning, clean-up, or stranding to remove the oil from the water. However, it does provide a good starting point from which to gain some insights concerning the amount of oil necessary to consume all of the oxygen in the waters of the Gulf of Mexico deeper than 200 m.

Since ZoBell's ratio of seawater to oil is approximately $1:10^6$ and if the volume of the Gulf of Mexico is assumed at approximately 2.352 M km^3 with an average oxygen concentration of 5 $mL \cdot L^{-1}$, it would require a spill of 6 trillion liters (31 billion barrels) to consume every atom of dissolved oxygen in that volume. This is equivalent to approximately the cargo of 63 thousand oil tankers. For perspective, the world's estimated crude oil reserves are ~1.2 trillion barrels (Energy Information Administration 2004), the U.S. reserves are ~23 billion barrels (Energy Information Administration 2002), and Gulf of Mexico U.S. reserves are ~5.2 billion barrels (Crawford et al. 2003). Thus, it is not possible that any spill in the Gulf of Mexico will completely oxidize the resident waters. However, locally and short term, the effect of an oil spill on oxygen concentration has the potential to be significant.

Two weeks after the Amoco Cadiz tanker spill (March 1978; South-Western Channel on the Portsall Rocks, three miles off the coast of Brittany, France; 223,000 metric tons) biodegradation rates were estimated to be between 0.01 and 0.027 mg oil per liter per year. At that time after the spill, the reduction of dissolved oxygen saturation in surface waters downstream of the spill ranged from 1 to 4.6%. The only possible explanation of the observed deficits is biodegradation (Aminot 1979).

Effects of discharge of produced waters: Produced waters are waters that are produced during normal oil and gas drilling and production operations. Currently, produced waters can be discharged with an average concentration 29 mg oil and grease L^{-1} per month. Using ZoBell's oxidation rate and *assuming no replenishment or ventilation of oxygenated waters*, the amount of produced water that would be needed to draw down the dissolved oxygen content in the water column below 800 m by 10 percent is estimated. If the total volume of the Gulf of Mexico below 800 m is $16x10^6$ km^3 then it would take approximately $16x10^{11}$ L of oil to reduce the dissolved oxygen by 10 percent. Given the allowable discharge concentration, to discharge this quantity of oil would require discharge of approximately $8x10^{13}$ L of produced water. If it is assumed that the average Gulf of Mexico produced waters discharge rate is 250 ML of fluid per day (NRC 2003), it would take roughly 320 thousand days or 876 years to discharge the required amount of produced water to reduce the dissolved oxygen content in the water column below 800 m by 10 percent.

Content of Oxygen Reservoirs: The dissolved oxygen content of the Gulf of Mexico at specified depth layers is presented in Table 4.8. The layers are based on hydrographic (water mass) properties found in the Gulf of Mexico basin and the depth layers that define the sills of the two ports into the Gulf. A clear oxygen minimum is seen in the 200 to 800-m box. This minimum is associated with the Tropical Atlantic Central Water entering the Gulf through the Yucatan Channel (see Section 4.1.2 for details). The rationale for the delineation of the Gulf into these layers is more fully discussed in Chapter 5.1 when the Box Model is presented.

Table 4.8

Reservoirs and Their Dissolved Oxygen Content per km^2

Reservoir	Depth Range (m)	Thickness (m)	Volume per km^2 (TL*·km^{-2})	Average O_2 (mL·L^{-1})	O_2 Content per km^2 (Mmol-O_2 ·km^{-2})
1	0-200	200	0.2	4.18	37.4
2	200-800	600	0.6	2.99	80.2
3	800-1500	700	0.7	4.15	130
4	1500-2000	500	0.5	4.89	109
5	2000-4000	2000	2.0	5.01	448

*1 km^3 = 10^9 m^3, 1 m^3 = 10^3 L, so 1 km^3 = 10^{12} L = 1 TL
At NTP (0°C) 1 mg O_2 = 0.7 mL, 1 mg O_2 = 62.5 μmol O_2 so 1 mL O_2 = 44.7 μmol O_2

Simple model of impact: To further investigate the impact of an oil spill on water column oxygen concentration, a first-order estimate of oxidation in an idealized system is developed. The approach presented utilizes a simple box model. This model is designed to focus on the potential effect on dissolved oxygen concentrations of major discharges of oil to the marine environment.

The objective of the following calculation is to present a simple system that can be used to make estimates of the quantities of petroleum that would have to be oxidized in order to produce specified changes in the oxygen content in closed reservoirs (boxes) of specified size and initial oxygen concentration.

The dissolved oxygen distributions in the Gulf of Mexico for which the average oxygen content can be calculated define the reservoir depth ranges (Table 4.8). The reservoir volumes are determined by thickness (depth range) and planar size (km^2). From their average oxygen content, a concentration of dissolved O_2 in megamoles km^{-2} can be calculated. From this and the planar size of the reservoir, the total amount of oxygen can then be calculated. This approach allows one to scale calculations from an immediate area of influence caused by the injection of petroleum to a basin-wide estimate. The assumptions are that the reservoir behaves as a box that is closed (no fluxes in or out) and no processes other than the oxidation of petroleum (carbon to carbon dioxide) occur within the box. Obviously both of these assumptions are false, so this calculation should be regarded as an upper limit.

A central consideration is how to connect a volume of added petroleum (and/or hydrocarbon gases) to oxygen consumption. This involves two factors that are the number of moles (λ) of oxygen needed to oxidize one mole of petroleum carbon (pet-C) and the number of moles of pet-C per unit volume (V) of petroleum (C_{pet-C}).

$$pet-C + \lambda O_2 \rightarrow CO_2 + H_2O$$

The number of moles of oxygen consumed is then:

$$\Delta O_2 = \lambda\ C_{pet-C}V.$$

For simple sugar (CH_2O) λ is 1 and for methane (CH_4) λ is 2. Here the preliminary estimate of λ = 1.6 is used.

In sugar C is 40 wt.% and in methane C is 75 wt. %. Kinghorn (1983) gives typical ranges of wt.% C in petroleum of 84 to 87% and specific gravity of from 0.81 to 0.985 g cm^{-3}. Mid-range values of 85 wt. % C and 900 g L^{-1} yield 765 g pet-C L^{-1} = 64 mol pet-C L^{-1}. Combining this with λ gives a value of about 102 mol-O_2 consumed per liter of petroleum. Petroleum volume is often quoted in barrels. 1 "US petroleum barrel" is 42 gallons or (x 4.55 L $gallon^{-1}$) 191 L. Therefore, about 2 x10^4 mol-O_2 (0.02 Mmol) are consumed per barrel of petroleum. Based on the O_2 reservoirs given in Table 4.8, the number of barrels of petroleum per km^2 area necessary to produce 1) a 10% change in O_2, 2) hypoxia (2 mL O_2 L^{-1}), and 3) anoxia (complete O_2 consumption) have been calculated (Table 4.9). [Note that 100 barrels of oil is equivalent to a layer of oil 0.5 mm thick over 1 km^2.]

Table 4.9

Number of Barrels of Petroleum per km^2 to Produce a Given Change in Different Reservoirs

Reservoir	10% Loss	Hypoxia	Anoxia
1	187	975	1,870
2	401	1,320	4,010
3	650	3,360	6,500
4	505	3,220	5,050
5	2,240	13,450	22,400

The following points should be considered when investigating this model. Our approach is very simplistic and has no real world analogy. Essentially all the model does is titrate a volume of oxygenated water with oil derived – carbon producing an upper limit for O_2 consumption in a closed system. In fact, any closed basin with no exchange will go anoxic if there is enough labile carbon and time. The mitigating circumstances in the offshore Gulf of Mexico are many and include:

• First, the oil will be found as a surface slick with only the oil at the oil/water interface subject to degradation. Even for a spill at depth, it is unlikely that significant mixing of oil-in-water occurs unless there are very turbulent conditions. So the titration would be limited by the rate of oxidation across the liquid/liquid interface.

• Second, the losses to evaporation would be significant in the first hours of a spill reducing surface area and lability (see fourth below).

• Third, dissolution of components of the oil is very sparing, limiting actual physical transport into the water volume – you have the titrant and the receptor but they have to be brought physically together.

• Fourth, the microbial community needs to be "turned on". While hydrocarbon-oxidizing ability is present, the microbial community is dormant. Some field experiments suggest it takes days to weeks to induce exponential growth in the indigenous microbial community, by that time physical mixing and other processes have completely altered the hydrocarbons and moved them around significantly.

- Fifth, if the rate of bio-oxidative removal is similar to the rate of oxygen production or transport (advection/diffusion) – one could titrate forever with the greater Gulf essentially supplying an unlimited amount of oxygen.

- Sixth, nutrients may well be limiting, N and P in particular. Therefore, complete oxidation is unlikely unless there is a source of nutrients. Carbon far exceeds the other essential elements (including micronutrients such as iron).

- Seventh, the thicker the oil slick, the more it will damp wave activity and retard oxygen diffusion across the sea-air interface, but the slick itself will not be uniformly thick so this will vary considerably. Also, because it is the oil at the interface that affects the oxygen concentration (see 1 above), a thick slick of a given volume will cover less area than a thinner one of the same volume, thus bacterial activity, and hence oxygen reduction, will take longer.

As a final computation, Table 4.10 estimates the upper limit of dissolved oxygen consumption if all the hydrocarbon discharged from the various natural and anthropogenic sources was oxidized. The estimated maximum oxygen consumed then is compared to the total oxygen in selected volumes within the Gulf at an assumed average concentration as given in Table 4.11. To determine possible basin scale catastrophic effects from total O_2 consumption by the various sources (Table 4.10) on the total O_2 available in the selected volumes requires several major assumptions. First, the volume must be assumed to be closed, with no replenishment from the atmosphere, photosynthesis, or advection, and the hydrocarbon must be well mixed within the volume. Second, all the hydrocarbon from the selected source must be confined within the selected volume. Third, all the hydrocarbon must be labile, and it must be fully oxidized within the selected water volume. These assumptions are not realistic, but do allow a zeroth order comparison to determine if the effects from the anthropogenic sources could be substantial on a basin scale.

The comparison itself shows that basin-scale effects from any of the annual discharges are negligible in all of the specified volumes, including the oxygen minimum zone and the deep Gulf waters below 1500 m. Even IXTOC-like spills would have minimal effects, as such a spill would consume at most only 1/10,000[th] of the available O_2 in the selected volumes. For example compare the total O_2 of 1.27E7 Mmol-O_2 in the oxygen minimum zone in the southeast Gulf with the total O_2 of 1.29E3 Mmol-O_2 consumed for an IXTOC-sized spill fully oxidized within and totally confined to that volume.

These computations make the major point that anthropogenic activities, even those of catastrophic spills, will not have major basin scale effects on the levels of dissolved oxygen in the Gulf of Mexico. The reader is reminded that the focus here has been on oxygen consumption, and so this conclusion may not apply to other environmental effects that may result from anthropogenic activities. Any major effects on dissolved oxygen from these activities will be localized, where they could be substantial within a few kilometers of the source. Neither the simple models in this subsection nor the simple box model of Section 5 were designed to address such localized effects.

Summary: The "worst" case computations and the simple model show that the catastrophic draw down of dissolved oxygen in the deep region of the Gulf of Mexico would not occur without a breakdown of the earth's thermohaline circulation, which provides the ventilation of dissolved oxygen to the deep waters in the Gulf. The ventilation and replenishment processes in the Gulf of Mexico prevent basin wide effects. However, the impact on dissolved oxygen concentrations of hydrocarbon inputs can be local. Determination of the nature and extent of such local effects depends on a complex system of interactions between the chemical nature and fate of the discharged material, the rates of dissolved oxygen replenishment at the site, and the environmental conditions existing at the discharge point and within the region of impact of the discharge plume. To estimate these local effects requires specific information on the discharge and its environmental setting and/or a complex modeling effort to tie together the many physical, chemical, biological, and geological factors that control the processes of degradation of hydrocarbons in the ocean.

Table 4.10

Dissolved Oxygen Consumed by Total Oxidation of Selected Hydrocarbon Sources

Hydrocarbon Source	Hydrocarbon Quantity (10^6 barrels[a])	Oxygen Consumed (Mmol-O_2)[b]
Natural Seeps[1]	0.98	378
Extraction-Produced Waters[1]	0.0119	4.59
Extraction-Other[1]	0.0021	0.81
Transportation[1]	0.0112	4.32
Consumption[1]	0.0476	18.38
IXTOC I spill[2]	3.332	1,286
U.S. Gulf of Mexico Proved & Unproved Reserves[3]	5,225	2.02E+06
U.S. Proved Reserves[4]	22,677	8.75E+06

[1] NRC 2003
[2] NRC 1985, 2003
[3] Crawford et al. 2003
[4] Energy Information Administration 2002
[a] assumes 1 tonne = 294 gallons and 1 barrel = 42 gallons
[b] assumes 1 liter of petroleum oxidizes 102 mol-O_2; 1 Mmol = 10^6 mol

Table 4.11

Dissolved Oxygen Content by Reservoir
(Whole Gulf is only that part of the Gulf of Mexico in water depths of 200 m or greater, i.e., the shelf is excluded.)

Reservoir Description	Depth Range (m)	Volume (10^6 km^3)	Oxygen Concentration (mL·L^{-1})	Oxygen Content (10^6 Mmol-O$_2$)[a]
Whole Gulf				
Upper Layer	0-200	0.239	4.18	44.7
Oxygen Minimum	200-800	0.557	2.99	74.4
Between Sills	800-1500	0.549	4.15	102
Deep Layer	≥ 1500	1.007	5.00	225
Surface to Bottom	0-bottom	2.352	4.97	522
Oxygen Minimum				
Southeast Gulf	200-800	0.095	2.99	12.7
Northeast Gulf	200-800	0.129	2.99	17.2
Northwest Gulf	200-800	0.113	2.99	15.1
Southwest Gulf	200-800	0.214	2.99	28.6
Deep Layer				
Southeast Gulf	≥ 1500	0.136	5.00	30.4
Northeast Gulf	≥ 1500	0.243	5.00	54.4
Northwest Gulf	≥ 1500	0.142	5.00	31.8
Southwest Gulf	≥ 1500	0.482	5.00	107.8

[a] assumes 1 mL O$_2$ = 44.7 μmol-O$_2$

5 BOX MODEL

A box model was constructed that modeled the complexity of the distribution of oxygen in the deep water region of the Gulf of Mexico. The model, despite taking a simple approach to a complex problem, provides useful insights into how the distribution of dissolved oxygen in the deepwater Gulf might be established and maintained. Among those insights are the following key findings on the importance of transport, reactive loss, and primary productivity:

1. It is highly probable that the oxygen profiles in the Gulf are determined by a *unique* combination of the transport process associated with circulation and mixing and the reactive processes associated with biogeochemical cycles. It is unlikely that a multitude of oceanographic and biogeochemical states could describe the observed oxygen profiles.

2. The transport through the Yucatan Channel plays the most significant role in controlling the oxygen profiles in the deep waters of the Gulf. The model suggests that approximately 30% of the Yucatan flow is diverted into the Gulf, although this may be a model artifact. Unfortunately the model is unable to provide an independent estimate of the flux necessary to ventilate the deep Gulf waters below the 2000-m sill depth of the Yucatan Channel. This vertical mass flux is critical to maintaining the deep oxygen profiles. But, because the actual mechanisms for transport of deep waters into the Gulf interior and for vertical mixing and ventilation of waters below the Yucatan sill depth are not well known, the necessary vertical flux of mass cannot be resolved by the box model.

3. The oxidation of carbon in the water column plays a secondary, but important, role to that of transport in maintaining oxygen profiles in the upper 800 m of the Gulf. The model is incapable of ascertaining its importance below 800 m.

4. The oxidation at the sediment interface plays a minor, but necessary, role in maintaining the oxygen below 1500 m. The model is incapable of ascertaining its importance above 1500 m. By design, it has no effect in the top 200 m.

5. The net production of oxygen in the surface layer is consistent with the rates reported for air-sea exchanges and photosynthetic inputs at the latitude associated with the Gulf.

6. The oxygen content in the Gulf would decrease to half of its present value in 10 years were the Gulf to be isolated from any source of oxygen, be it transport, primary productivity, or atmospheric exchange.

These insights should provide the basis for specifying the additional observations and modeling needed to more accurately understand the processes that control the distribution of dissolved oxygen in the deepwater Gulf of Mexico.

5.1   Introduction

The goal of the "Simple Box Model" Task is to construct a box model that adequately represents the inputs to and outputs from the deepwater Gulf of Mexico in a manner that reflects the complexity of the distribution of oxygen. In essence, the focus of the model is to gain insights into how the distribution of dissolved oxygen in the deepwater Gulf is established and apparently maintained over time and space, thus providing an improved basis for specification of observations needed to more accurately determine/predict changes in the oxygen content and structure of the basin.

Our hypothesis is that the distribution of oxygen within the water column is maintained by a combination of two major processes: the transport processes associated with circulation and mixing, and the reactive processes associated with biogeochemical cycles. This simple box model consists of five elements that will allow us to address how the distribution might be maintained:

1. Vertical and horizontal spatial definition of the boxes.
2. Estimates of the mass transport between and through boxes, in particular how much of the Loop Current flow must be diverted into the interior of the Gulf.
3. The sources and sinks of oxygen consisting of
    a). Physical – atmospheric and water masses
    b). Biological – photosynthesis and respiration
    c). Chemical – abiotic oxygen consumption
    d). Geological – hydrocarbon seeps and vents
    e). Anthropogenic – hydrocarbon releases
4. The governing equations describing the accumulation, transport, and reaction of oxygen.
5. The inclusion of conservative tracers, i.e., temperature and salinity, to close the mass transport balance.

The level of detail chosen to describe each of the five elements defines the complexity of the model. It is highly dependent on the availability of data and understanding of the physical and biogeochemical processes active within the Gulf. As stated in Section 4, the data available to determine flux rates are sparse and knowledge of the circulation in the deep Gulf below about 1000 m is poor. Thus, this box model is limited to essentially zero-order approximations. Nonetheless, we believe that it provides insights into the gaps in knowledge and how they might be resolved (see Section 6).

The model setup, as well as the underlying assumptions of a box model, are discussed in Section 5.2. Model equations are defined in Section 5.3. The results of the baseline model run are given in Section 5.4. Several scenarios are presented in Section 5.5. The conclusions are in Section 5.6.

5.2   The Model Setup

The model setup section addresses the first three elements outlined in the introduction. These are discussed in the following three subsections. Section 5.2.1 addresses various aspects of the vertical and horizontal spatial definition of the boxes, Section 5.2.2 estimates the mass transport between and through the boxes, and Section 5.2.3 discusses the sources and sinks of oxygen relative to the box model.

5.2.1   Definition of the Boxes

A box model requires that the spatial extent of the individual boxes be defined. These are defined in terms of horizontal and vertical sizes, which in this box model are defined as

> *a. Horizontal Box Size:* There are insufficient high quality oxygen data, extending below 2000 m, to allow realistic division of the Gulf into east and west boxes. Thus, a single basin model is posed that has no western or eastern basin, but encompasses the entire Gulf. In essence, the mechanisms that maintain a standard dissolved oxygen profile will be examined. Hence, only the vertical division of boxes must be defined.

> *b. Vertical Box Size:* The available bottle dissolved oxygen data taken in the Yucatan channel are shown versus depth in Figure 5.1. The general vertical structure in the channel, as well as numerous profiles taken across the Gulf, show similar regions of recognizable trends in the oxygen distribution. Based on these trends, the single basin model is subdivided into five vertical layers (see Figure 5.2) that are intended to mirror the expected biological (reactive processes) and physical processes (transport processes). The box model was based on these five vertical layers. The layers are shown schematically in Figure 5.3 along with presumed transport pathways. Details of each box, including volume and surface area, are presented in Table 5.1.

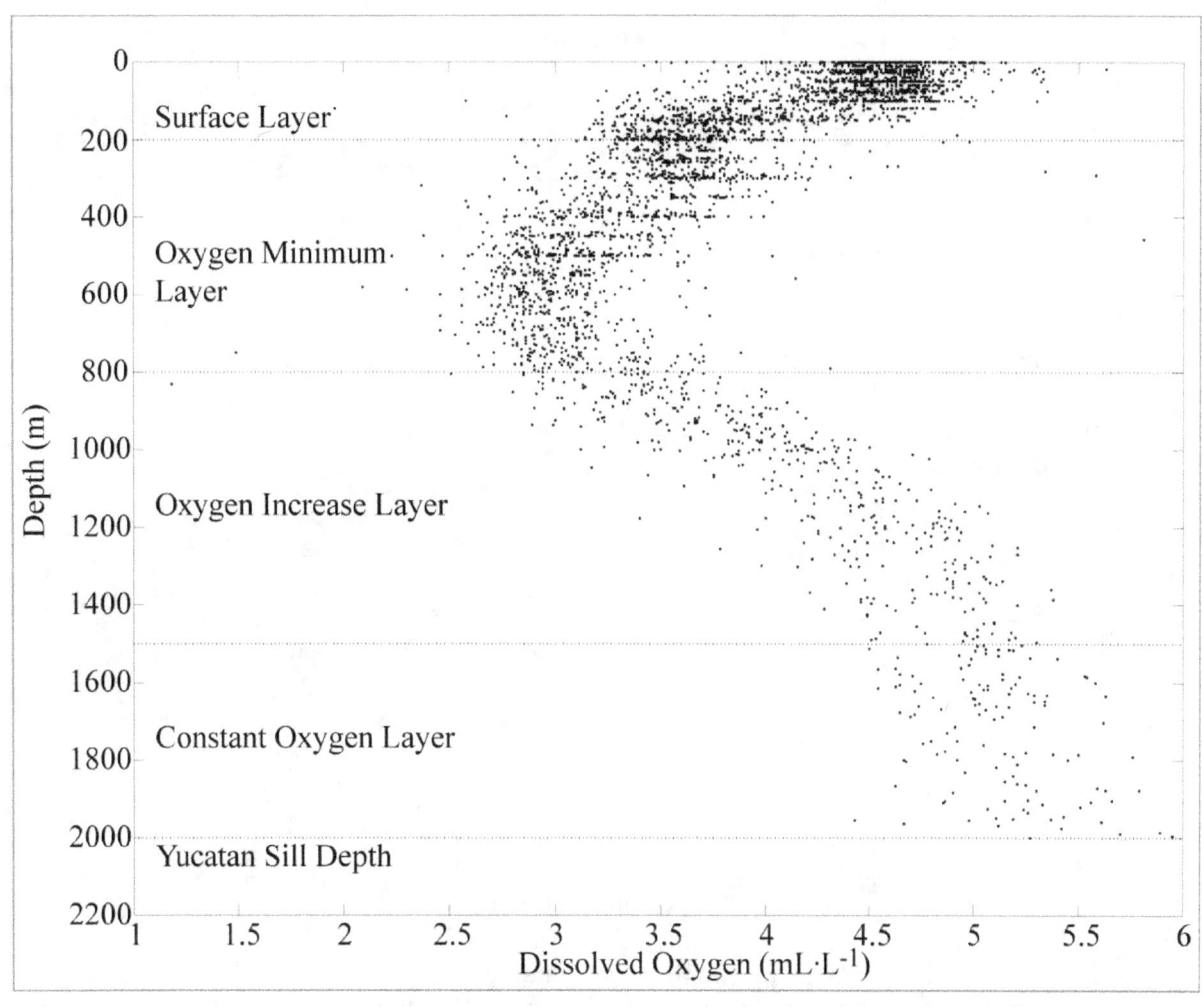

Figure 5.1. Dissolved oxygen versus depth for the available data in the Yucatan Channel.

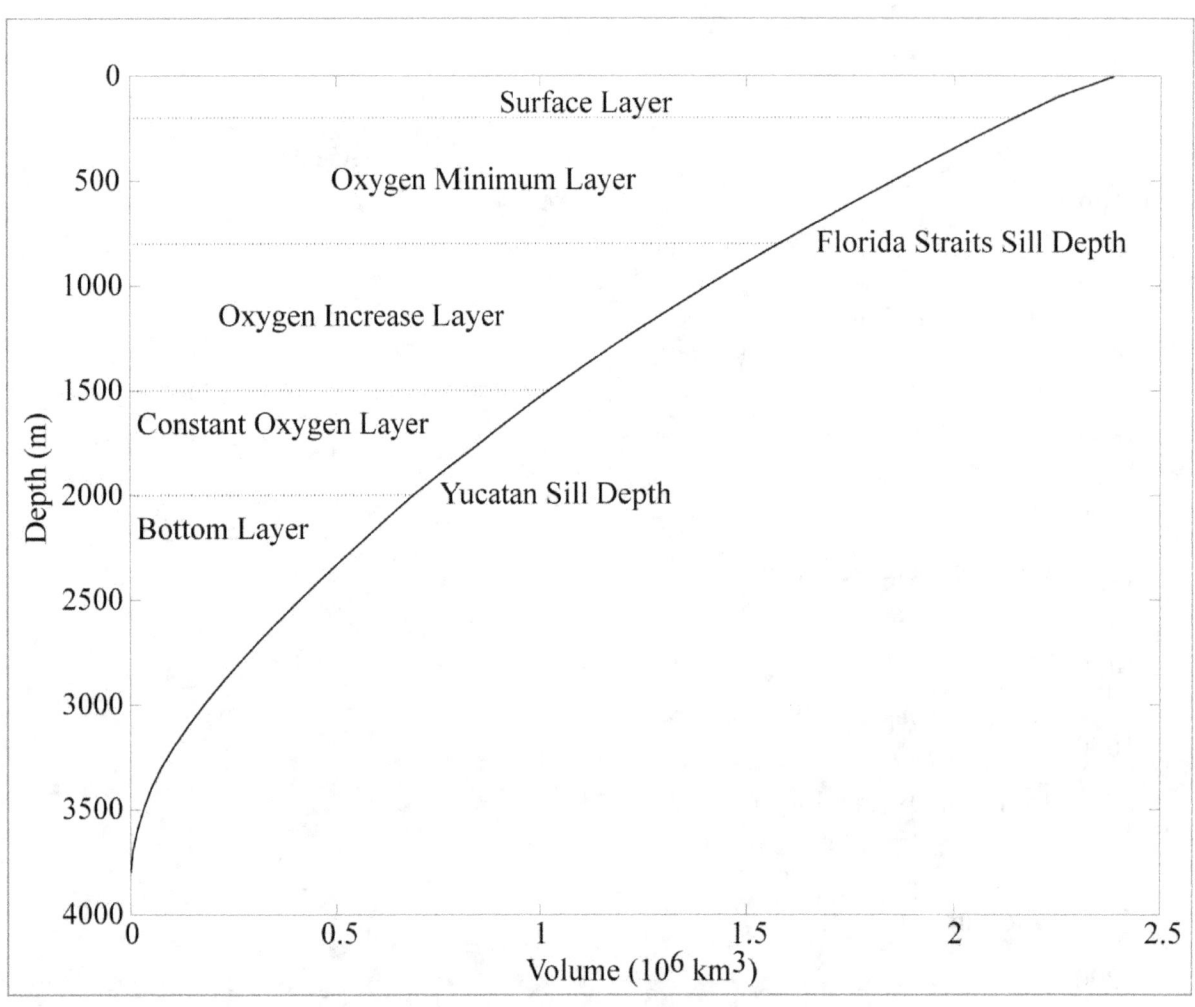

Figure 5.2. Cumulative volume by depth of the deepwater Gulf of Mexico and the vertical delineation between layers. Details about the volume of each layer and the associated area of the sediment interface are provided in Table 5.1.

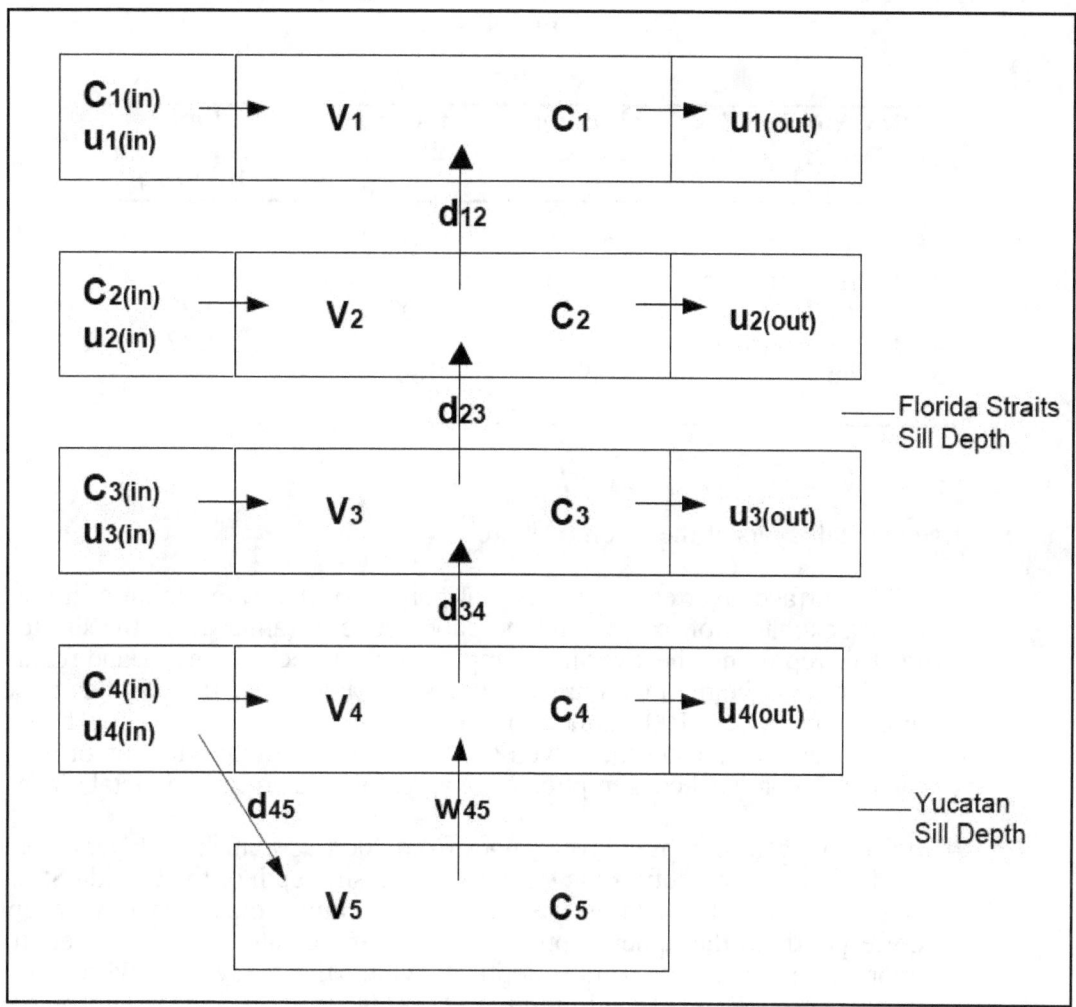

Figure 5.3. Box model configuration and notation used for the dissolved oxygen mass balance. The arrows represent fluxes. An option is included to allow for a deep return flow through the Yucatan Channel, i.e., u_4(out). Below the 2000-m sill depth of the Yucatan Channel, there is no horizontal flux into or out of the box.

Table 5.1

Volume, Thickness, and Sediment Surface Areas for Each Layer in the Basin-wide Box Model of the Gulf of Mexico

Box Number & Description	Volume (10^5 km^3)	Layer Thickness (m)	Sediment Surface Area (10^3 km^2)
1. Surface	2.040	200	-
2. Oxygen Minimum	5.653	600	3.085
3. Oxygen Increase	5.559	700	4.423
4. Oxygen Constant	3.300	500	2.242
5. Bottom	6.922	1800	5.874

The five vertical layers of the box model are

i. The surface layer encompasses that portion of the water column influenced by air-sea exchange of oxygen and the photic zone. It ranges from the surface to 200 m. This represents the layer most impacted by surface processes and features such as winds and the thermocline. It includes the photic zone that extends to depths of approximately 60-100 m. It is dominated by the flux of oxygen between the atmosphere and the ocean, as well as the photosynthetic production of oxygen and the rapid biological consumption of oxygen by organic carbon metabolism.

ii. The oxygen-minimum layer ranges from 200 to 800 m. This layer extends from the bottom of the photic zone to the accepted sill depth of the Florida Straits. This is the region that contains the oxygen minimum zone. The lower depth also corresponds to the upper depths of the Antarctic Intermediate Water. It is also approximately the maximum depth to which the isopycnals within a ring are depressed. The oxygen minimum at 400-600 m is in part a relic of processes occurring upstream; this layer also is primarily a region of oxygen consumption.

iii. The oxygen-increase layer ranges from 800 to 1500 m, from the sill depth of the Florida Straits to the approximate depth at which dissolved oxygen no longer increases significantly in concentration with vertical depth.

iv. The constant oxygen layer ranges from 1500 to 2000 m, the approximate sill depth of the Yucatan Channel. This is a region of maximum oxygen concentration below the surface layer.

v. The bottom layer ranges from 2000 m to 3800 m, i.e., from the Yucatan Channel sill depth to the deepest depth of the Gulf of Mexico. This is a region where the dissolved oxygen concentration changes very little with increasing depth.

c. Well-Mixed Volumes for Each Box: The implied assumption in any box model is that the volume in each box is so well mixed that any changes are instantly communicated to the entire box. Thus, the boxes should be made small enough relative to the processes being studied so that this assumption is not grossly violated.

d. Rigid-Lid Approximation: Another implied assumption in this box model is that the volume in each box must remain constant over time. This constraint implies that either the time scale is long enough so that temporal changes in the surface elevation average to zero or the volume changes associated with the surface elevation variations are so small compared to the bathymetric volume of the Gulf that they can be neglected. Both conditions hold. The bathymetric volume of the Gulf, estimated from a standard bathymetry data set, is 2.352×10^6 km^3. The corresponding surface area is 1.631×10^6 km^2. A 1% change in the volume of the Gulf would correspond to a 15-m change in the surface elevation. Changes in the surface elevation are the result of changes in the Loop Current transport, tides, the river input, precipitation, and evaporation. Altimetry from the TOPEX/Poseidon and ERS-2 satellites show the deviation in surface elevation across the Gulf rarely exceeds ±1 m, or no more than a 0.07% change in total volume. Consequently, the rigid lid approximation holds.

e. Exclusion of the Shelf: In this box model, the shelf is not included and the deep ocean exchange with and processes associated with the shelf are neglected. This assumption is based primarily on the small volume of the shelf relative to the remainder of the Gulf. Only 8.5% of the Gulf's volume occurs in depths from the surface to 200 m and, of that volume, 15.8% occurs on the shelf. It is assumed that the contribution of processes over the shelf and any exchanges from land to the dissolved oxygen concentrations in the deep Gulf interior can be reasonably neglected. Furthermore, the contribution from the shelf will be implicitly included in the oxygen concentration initial conditions, to be discussed later. Excluding the shelf is a reasonable assumption given the crude resolution of this box model and the knowledge that the continental slope acts as an effective barrier to transport across the shelf.

5.2.2 Mass Transport

A box model makes no specific distinction as to the source of the inflow, the destination of the outflow, or how much vertical transport there has to be. It essentially integrates the individual details into a net transport. To make the model physically realistic, the major inflow, outflow, and vertical fluxes must be judiciously specified.

a. *Inflow Estimates:* The major inflow into the box model, designated $u_1(in)$, $u_2(in)$, $u_3(in)$, and $u_4(in)$, (see Figure 5.3) is assumed to be diverted from the Loop Current. Thus, the question is: exactly how much is diverted. Altimeter fields from the TOPEX/Poseidon and ERS-2 satellites show clear indications that the Loop Current exchanges fluid with the interior of the Gulf of Mexico. In this model, that exchange is parameterized as the unknown β, a number that defines the fraction of the total Loop Current flow diverted into the interior. It will be shown that the parameter β can be reasonably constrained by the physical restrictions inherent in this model.

The total flux through the Yucatan Channel is fairly well known, but the vertical distribution of that flux, averaged across the channel, is not. To avoid the need for a reliable and extensive current meter data set in the Yucatan to accurately define the vertical distribution, we take two approaches. In the first, the horizontal flux is apportioned in a systematic and reproducible way using a polynomial velocity profile of the simple form

$$u(z) = mz^n$$

The velocity is a maximum at the surface and is zero at the bottom, This simplified profile is assumed to be adequate for the purposes of this zero order box model. The coefficient m is determined by the total flux

$$q_{total} = \frac{m}{n+1} H^{n+1}$$

where H is the total depth and the total flux is specified.

In the second approach the work of Sheinbaum et al. (2002) is used to make the flux estimates. They made direct current observations across the Yucatan Channel for a period of ten months and presented the structure of the mean along-channel velocity field as their Figure 2a. In the absence of the actual data, a graphical representation of the flux profile was extracted from their figure. For the case of a Yucatan Channel inflow of 30 Sv and a sill depth of 2000 m, the profiles for various choices of n as well as the extracted Sheinbaum profile, are shown in Figure 5.4. Table 5.2 lists the actual flux values pertinent to each box. Increasing the order of the polynomial n effectively forces more of the flux to occur closer to the surface.

b. *Outflow Estimates*: The major outflow from the box model rejoins the Loop Current before exiting the Gulf through the Florida Straits. The manner in which this box model is designed, coupled with the continuity constraint, ensures that the outflow is a dependent variable. Consequently, if 30 Sv enters the Gulf through the Yucatan Channel, then 30 Sv must exit. Part of that 30 Sv is diverted into the Gulf interior, the realm of this box model, and if the water level in the Gulf is to remain constant over a long period of time, it leaves the interior of the Gulf and reenters the Loop Current. We should also note that the 800-m sill depth of the Florida Straits effectively guarantees that only boxes 1 and 2 can have this outflow, i.e., $u_1(out)$ and $u_2(out)$. But because there is evidence (Bunge et al. 2002) of a deep return flow to the Caribbean through the Yucatan Channel, an outflow is included for boxes 3 and 4, $u_3(out)$ and $u_4(out)$, to allow for a return flow that is a box model unknown.

c. *Vertical Transports*: Because the vertical velocities are much smaller than the horizontal velocities, the vertical flux of mass in the ocean is extremely difficult to accurately measure. For the purposes of this model, the vertical flux could be roughly estimated from the hydrostatic continuity balance. Originally we had anticipated doing just this, but quickly discovered this approach can be quite misleading if the scaled parameters are incorrectly estimated. To avoid this problem, the transport between the upper three boxes was determined as an output from the box model and are graphically shown in Figure 5.3 as d_{12}, d_{23}, and d_{34}. Because of the difference in sill depth between the Yucatan and the Florida Straits, and the small Yucatan return flow, continuity requires that the majority of the inflow to boxes 3 and 4 must be diverted vertically. The vertical transport, w_{45}, is the only means by which box 5 can be ventilated.

5.2.3   Sources and Sinks of Dissolved Oxygen

a. *Physical:* Each of the boxes will have, as a source and a sink, the physical transport of oxygen into and out of the box. Transport alone, however, is insufficient to maintain the distribution of oxygen within the water column.

b. *Photosynthesis and Atmospheric Transfer:* The rate of photosynthesis within the surface layer and the transfer of atmospheric oxygen into the surface layer are difficult to independently quantify from the literature. Consequently, the combined photosynthetic production and net atmospheric input of oxygen into the surface layer are determined as an output from the box model.

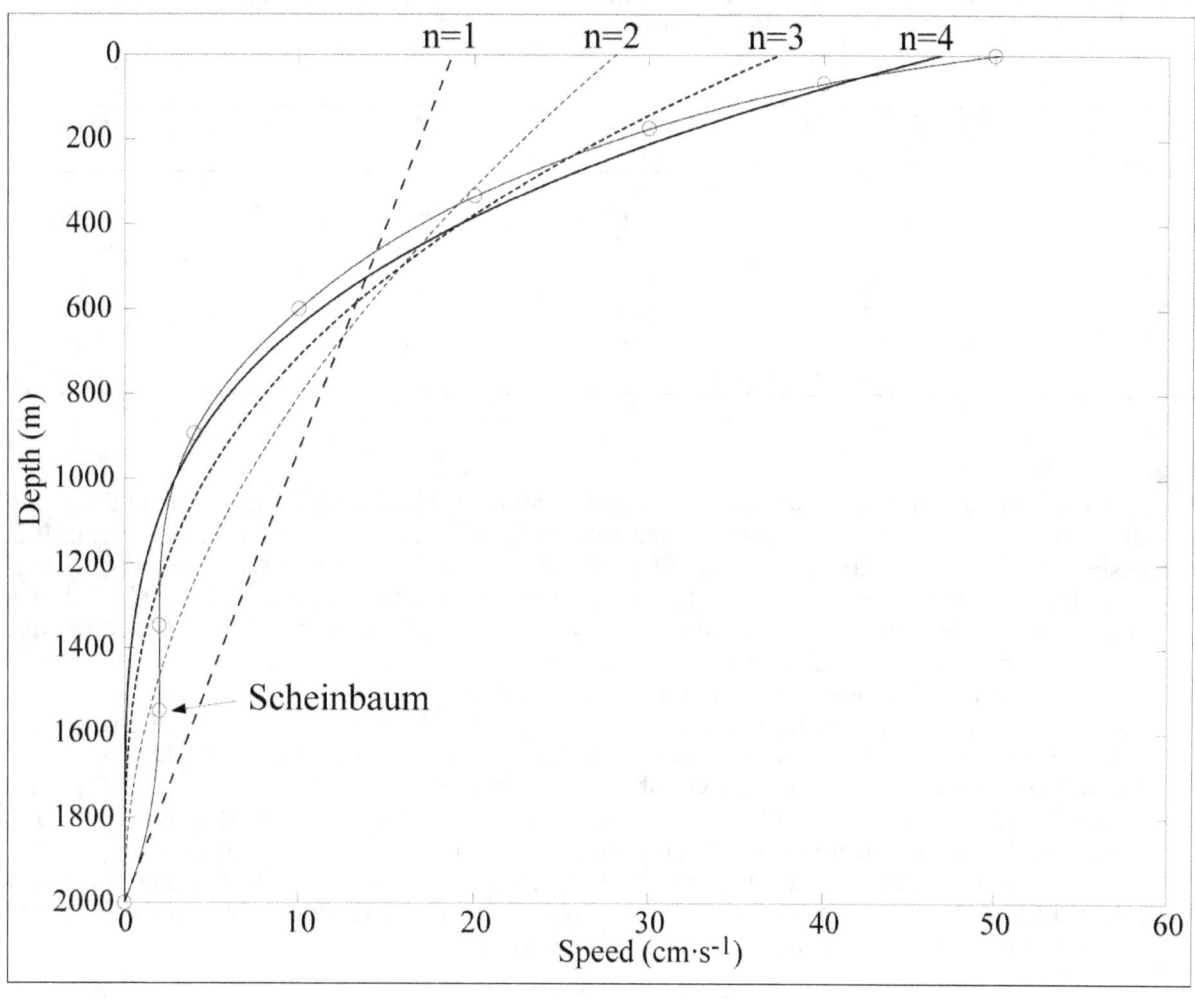

Figure 5.4. Velocity profiles of the Yucatan Channel inflow. Shown are the four cases of an assumed polynomial form, $u(z) \sim z^n$, and an estimate based on the direct current measurements across the Yucatan Channel made by Sheinbaum et al. (2002).

95

Table 5.2

Fluxes of the Yucatan Channel Inflow
(Fluxes in Sv are through boxes 1-4 of the model from the Yucatan Channel (YC) under selected
choices of polynomial order.)

Flux (Sv)	Sheinbaum et al. (2002)	n=2	n=3	n=4 -	n=5
YC_1	12.01	8.13	10.32	12.28	14.06
YC_2	13.72	15.39	15.80	15.38	14.54
YC_3	3.03	6.01	3.77	2.30	1.39
YC_4	1.15	0.469	0.117	0.029	0.007
YC total	30	30	30	30	30

c. Biogeochemical Processes in the Water Column: The oxidation of particulate and dissolved organic carbon within the water column results in the consumption of oxygen. It is assumed that this is the primary pathway for the reactive loss of dissolved oxygen. It is modeled, not as a second order reaction in oxygen and carbon, i.e., $r = kV[O_2][C]$, but as a simple first order reaction in oxygen, i.e., $r = k_w V[O_2]$. Thus, all possible biogeochemical reactions are parameterized into a single reaction with one rate constant. The alternative of including all possible reactions is untenable in a box model, particularly in this case where many of the rates are unknown. Furthermore, a second order reaction would require a separate box model to track carbon, a task that is beyond the scope of this simple box model. To account for the fact that the availability of labile carbon does decrease rapidly with increasing depth and is probably quite refractive in the deep Gulf, the first order reaction is only included in the upper two boxes and selectively turned off in the remaining three deeper boxes. This selection was not entirely arbitrary as experimentation with the model quickly showed that if the reaction was included below 800 m, the demarcation between the second and third box, then reasonable results were not possible.

d. Biogeochemical Processes at the Sediment-Water Interface: The consumption of oxygen at the sediment surface is the combination of the transfer of oxygen from the bulk to the surface followed by a sediment-mediated reaction. The rate at which this occurs can be modeled as

$$V \frac{d[O_2]_{surface}}{dt} = Ak_m \left([O_2]_{bulk} - [O_2]_{surface} \right) - r[O_2]_{surface}$$

where V is the volume, A is the surface area, k_m is the mass transfer coefficient from bulk to surface, r is the rate of disappearance by the sediment-mediated reaction, $[O_2]_{bulk}$ is the known oxygen concentration in solution, and $[O_2]_{surface}$ is the unknown concentration at the surface. The consumption of oxygen can be either mass transport limited or reaction limited. If the sediment is porous and the oxygen has to diffuse into the pores before it can react, then the rate is probably mass transport limited. In the interests of simplicity, a mass transport limited rate is assumed for this model, so that

$$V \frac{d\left[O_2\right]_{surface}}{dt} = Ak_m \left(\left[O_2\right]_{bulk} - \left[O_2\right]_{surface} \right)$$

If it further is assumed that the surface concentration is a linear fraction of the bulk, then

$$V \frac{d\left[O_2\right]_{bulk}}{dt} = Ak_m \frac{(1-x)}{x} \left[O_2\right]_{bulk} = Ak_b \left[O_2\right]_{bulk}$$

This has the same mathematical form as a first order reaction. Consequently, there is no mathematical difference between the first order water column reaction discussed above and this mass transport limited process. In fact, it is no different than a first order reaction occurring at the sediment-water interface. The rate constant and the mass transfer coefficient are mathematically identical and no independent information can be extracted from this box model by including both water column and sediment processes together in a single box.

Because each box is ideally well mixed, all of the fluid by definition must contact the sediment before it leaves the box. Unless the fluid spends a lot of time in the box, i.e., the overturning age is quite large, then complete contact with the sediment is unlikely. An option has been included to scale down the contribution expected from the sediment. To account for the fact that a water column reaction is imposed in the upper two boxes, it is only for the deepest two boxes that a sediment contribution can be included. Experimentation with the model resulted in determining the values that minimized the error of the model.

e. Geological and Anthropogenic: The presence of natural hydrocarbon seeps and vents and the inadvertent introduction of hydrocarbon from human activities into the Gulf are a potential sink for oxygen. Because the model only allows for a first order reactive loss of oxygen and does not independently follow the fate of carbon, this sink cannot be modeled directly. However, it can be estimated by an assumed increase in the rate constant of k_w, an approach attempted in the scenarios section.

5.3 The Model Equations

Before establishing the equations that will be used to mathematically model the distribution of oxygen in the Gulf of Mexico, key points from Section 5.2 are summarized. The distribution of oxygen within the water column is controlled by a combination of transport processes associated with circulation and mixing and by reactive processes associated with major biogeochemical cycles occurring in the water column and across the water-sediment interface. The box model has five boxes in the vertical, as shown in Figure 5.3. The difference in the sill depths between the Yucatan and the Florida Straits will have a significant effect on the results. The governing equations can now be defined, the unknowns specified, and a solution attempted.

5.3.1 Governing Equations

The basic equation governing tracer properties in each layer of this box model is posed in terms of a simple balance between accumulation, transport, and reaction, represented here as:

$$V \frac{dC}{dt} = u_{in} C_{in} - u_{out} C - d(C) - r(C) \quad (5.1)$$

Here C is the tracer concentration, denoted somewhat generically because it is meant to represent any tracer, not necessarily just dissolved oxygen. The variable u_{in} is the horizontal flow into the box, i.e., the inflow that brings with it material of concentration C_{in}. The variable u_{out} is the horizontal flow out of the box that carries with it material of concentration C that is within the box. There is also a vertical flux either into or out of (but not both) the top of the box, as well as a vertical flux into or out of the bottom of the box. The variable $d(C)$ is meant to represent this vertical flux of material between overlying boxes. Details of how this is mathematically expressed will be presented below. The variable $r(C)$ represents the first order process, either a water column reaction, photosynthesis in the photic zone, or a sediment mediated reaction, that either produces or consumes oxygen. Equation (5.1) simply states that the accumulation of a tracer, VdC/dt, i.e., the time rate of change of the tracer in the box of volume V, is the result of two major processes: the transport processes associated with circulation and mixing, $u_{in} \cdot C_{in} - u_{out} \cdot C - d \cdot C$, and the reactive processes associated with biogeochemical cycles, $r(C)$. As a consequence of the well-mixed assumption, C is the concentration of the tracer everywhere within the box; consequently it is also the concentration of the tracer leaving the box. If the transport of incoming tracer C_{in} exceeds the loss by outflow and reaction, then the concentration in the box will grow in time, that is, until something changes. But, however interesting that evolution might be, time dependence is well beyond the scope of this simple box model. Steady state is assumed so that $VdC/dt = 0$.

The fate of oxygen is of primary interest to this study, and its fate is largely, but not completely, controlled by transport. Getting those transports correct, even within the context of this zero-order box model, is pivotal. A box model based on tracking oxygen as the sole tracer is entirely feasible, and was an approach we initially investigated. But this approach can be improved upon. Oxygen is not a conservative tracer; the reactive processes play an important role that must be accounted for by the box model. This complicates the box model in that both transport and reaction must be determined simultaneously if oxygen is used as the only tracer. If additional tracers can be utilized, tracers that are conservative and are simply transported with the fluid, then it is advantageous to make use of them in order to obtain a more reliable estimate of the physical transport. We decided to make use of the extensive set of temperature and salinity profiles in the Gulf as those conservative tracers.

The expansion of equation (5.1) for oxygen, temperature, and salinity gives the following set of equations for each of the five boxes:

Box 1

$Continuity:$ $\quad 0 = u_1^{in} - u_1^{out} + d_{12}$

$Salt \cdot balance:$ $\quad 0 = u_1^{in} S_1^{in} - u_1^{out} S_1 + d_{12}\{S_1, S_2\}$

$Heat \cdot balance:$ $\quad 0 = u_1^{in} T_1^{in} - u_1^{out} T_1 + d_{12}\{T_1, T_2\}$ $\hfill (5.2)$

$Oxygen \cdot balance:$ $\quad 0 = u_1^{in} O_1^{in} - u_1^{out} O_1 + d_{12}\{O_1, O_2\} - k_w V_1 O_1 + V_1 r_P(O_1)$

Box 2

$$Continuity: \quad 0 = u_2^{in} - u_2^{out} + d_{23} - d_{12}$$

$$Salt \cdot balance: \quad 0 = u_2^{in} S_2^{in} - u_2^{out} S_2 + d_{23}\{S_2, S_3\} - d_{12}\{S_1, S_2\}$$

$$Heat \cdot balance: \quad 0 = u_2^{in} T_2^{in} - u_2^{out} T_2 + d_{23}\{T_2, T_3\} - d_{12}\{T_1, T_2\}$$

$$Oxygen \cdot balance: \quad 0 = u_2^{in} O_2^{in} - u_2^{out} O_2 + d_{23}\{O_2, O_3\} - d_{12}\{O_1, O_2\} - V_2 k_w O_2$$

(5.3)

Box 3

$$Continuity: \quad 0 = u_3^{in} - u_3^{out} + d_{34} - d_{23}$$

$$Salt \cdot balance: \quad 0 = u_3^{in} S_3^{in} - u_3^{out} S_3 + d_{34}\{S_3, S_4\} - d_{23}\{S_2, S_3\}$$

$$Heat \cdot balance: \quad 0 = u_3^{in} T_3^{in} - u_3^{out} T_3 + d_{34}\{T_3, T_4\} - d_{23}\{T_2, T_3\}$$

$$Oxygen \cdot balance: \quad 0 = u_3^{in} O_3^{in} - u_3^{out} O_3 + d_{34}\{O_3, O_4\} - d_{23}\{O_2, O_3\} - \alpha_3 A_3 k_b O_3$$

(5.4)

Box 4

$$Continuity: \quad 0 = u_4^{in} - u_4^{out} - d_{34} + d_{45}$$

$$Salt \cdot balance: \quad 0 = u_4^{in} S_4^{in} - u_4^{out} S_4 - d_{34}\{S_3, S_4\} + d_{45} S_5$$

$$Heat \cdot balance: \quad 0 = u_4^{in} T_4^{in} - u_4^{out} T_4 - d_{34}\{T_3, T_4\} + d_{45} T_5$$

$$Oxygen \cdot balance: \quad 0 = u_4^{in} O_4^{in} - u_4^{out} O_4 - d_{34}\{O_3, O_4\} + d_{45} O_5 - \alpha_4 A_4 k_b O_4$$

(5.5)

Box 5

$$Continuity: \quad 0 = w_{45} - d_{45}$$

$$Salt \cdot balance: \quad 0 = w_{45} S_4^{in} - d_{45} S_5$$

$$Heat \cdot balance: \quad 0 = w_{45} T_4^{in} - d_{45} T_5$$

$$Oxygen \cdot balance: \quad 0 = w_{45} O_4^{in} - d_{45} O_5 - \alpha_5 A_5 k_b O_5$$

(5.6)

Here O_n is the dissolved oxygen concentration, S_n is the salinity, and T_n is the temperature within each box; V_n is the volume; A_n is the sediment surface area; u_n^{in} is the horizontal flux into each box and u_n^{out} is the horizontal flux out of each box; d_{12}, d_{23}, d_{34} and d_{45} are the vertical fluxes between boxes, k_w is the first order rate constant in the bulk, k_b is the mass transfer coefficient at the water-sediment interface; α_n is the scaling factor controlling the contribution of the sediment process, and $r_P(O_2)$ is the combined photosynthetic production of and net atmospheric input of oxygen into the surface layer.

For each box, there is a balance of mass transported into and out of the box, known as continuity. For the sake of consistent notation, both horizontal and vertical fluxes are designated positive if they flow into a box and negative if they flow out. All vertical fluxes are given as d_{ij}, which designates the flux between two adjacent boxes i and j. In addition to continuity, there is a tracer balance. A conservative tracer is swept along by the flow and is neither created nor destroyed. For the purposes of this model, we assume that salt and heat are ideal. Removed from the influences of the surface where freshwater can be added or the heat content increased by solar radiation, this assumption holds quite well. At the surface, we accept the small errors introduced by this assumption because on the overall scale of this box model those errors are negligible. The notation $d_{ij}\{C_i, C_j\}$ is our shorthand for the vertical transport of tracer C. The value of the transport depends on whether the unknown vertical flux d_{ij} flows up toward the surface, a positive quantity, or down toward the bottom, a negative quantity. For example, if d_{23} is positive, then it will carry tracer C_3 up to box 2, but if d_{23} is negative, then it will carry tracer C_2 down to box 3. The notation $d_{ij}\{C_i, C_j\}$ is meant to mathematically represent these possibilities, which the box model solves in an iterative manner.

As we have already noted, it is extremely difficult to determine a physical value for the vertical flux into or out of box 5. An examination of the equation set for box 5 shows that including salinity and temperature does not solve the problem. If the two salinities are in fact different, then the salt balance equation cannot be satisfied unless w_{45} is different from d_{45}. If the two temperatures are in fact different, then the heat balance equation cannot be satisfied unless w_{45} is different from d_{45}. But continuity clearly requires that w_{45} is equal to d_{45}. The oxygen balance can be satisfied only because it is nonconservative, i.e., the sediment mediated reaction consumes enough oxygen so that the two concentrations can be different. The box model solves for w_{45} as an unknown. This difficulty in constraining the vertical flux highlights a significant knowledge gap in determining how the distribution of oxygen within the water column is maintained, that of the vertical fluxes below the 2000-m depth.

5.3.2   Specification of Knowns and Unknowns

Considering the equation set (5.2 to 5.6), we could estimate values for the vertical fluxes and determine the resulting oxygen concentrations, to then be compared against the known values, or the known oxygen concentrations could be used to place a bound on the vertical fluxes. The latter approach is chosen and the unknowns are as follows:

1) The horizontal outflows from each box, $u_1(out)$, $u_2(out)$, $u_3(out)$, and $u_4(out)$;
2) The inflow into box 4, $u_4(in)$, and into box 5, w_{45}.
2) The vertical fluxes, d_{12}, d_{23}, d_{34}, and d_{45};
3) The production of oxygen in the photic zone by some combination of photosynthesis and atmospheric transfer, and;
4) The two rate constants controlling the reactive loss of oxygen in the water column, k_w, and at the sediment-water interface, k_b.

The knowns are the concentration of salinity, temperature, and oxygen and the horizontal inflows to the upper four boxes and are specified *a priori*. Table 5.3 shows the values of temperature, salinity, and oxygen used for the inflow through the Yucatan Channel and for the Gulf interior. Section 3.3 discusses how these values were obtained. The value used for each box was the mean value of the 1-m resolved data averaged over the depth of the box.

An examination of the oxygen data explains the rational for the design of flow into the box. Note specifically that the oxygen concentration in box 4, 4.95 mL L^{-1}, is actually less than that of box 5, 5.04 mL L^{-1}. The only reasonable means by which the concentration in box 5 (which receives no horizontal transport from the Yucatan Channel inflow because the box lies below the sill depth) could be higher than the overlying box is if there is an alternative, higher concentration source. The likely source is the 5.12 mL L^{-1} oxygen in the lowest box of the Yucatan. Perhaps the contention of Welsh and Inoue (2000) that Loop Current rings are the mechanism for driving the deep circulation is correct and eddies are in fact a major mechanism for transport of water to the

100

deep Gulf. In order to formulate a reasonable box model we had to include a mechanism for injecting high oxygen concentration fluid below 2000 m.

Table 5.3

Values of Temperature, Salinity, and Dissolved Oxygen Used for the Yucatan Channel Inflow and Gulf Interior

Box Number	Temperature (°C)	Salinity	Dissolved Oxygen ($mL \cdot L^{-1}$)
Yucatan Channel			
1	24.38	36.32	4.11
2	11.72	35.50	3.17
3	4.76	34.94	4.31
4	4.06	34.97	5.12
Gulf Interior			
1	20.74	36.23	3.87
2	9.27	35.20	3.02
3	4.60	34.94	4.42
4	4.07	34.97	4.95
5	4.00	34.97	5.04
Florida Straits			
1	22.68	36.25	3.97
2	10.88	35.38	3.08

5.3.3   Additional Details

To maintain a constant volume, the following set of continuity constraints must hold, beginning with the overall continuity balance:

$$u_1^{in} + u_2^{in} + u_3^{in} + u_4^{in} = u_1^{out} + u_2^{out} + u_3^{out} + u_4^{out} \tag{5.7}$$

The horizontal inflow to each box is determined by the fraction of the inflow from the Yucatan Channel that leaves the Loop Current and enters the Gulf interior. This fraction is the parameter, β, which can range from 0 to 1 and is determined by minimizing the error in the box model, where the error is yet to be defined. The equation set is:

$$u_1^{in} = \beta \cdot YC_1$$
$$u_2^{in} = \beta \cdot YC_2$$
$$u_3^{in} = YC_3 \qquad (5.8)$$
$$u_4^{in} + w_{45} = YC_4$$

It is recognized that the difference in sill depths between the Yucatan Channel and the Florida Straits forces all of the flow below 800 m to enter the Gulf interior. The implied assumption is that the Yucatan Channel return flux, YC_R, originates in the Gulf and is not a countercurrent to the Loop Current. In this model YC_R is determined as part of the box model.

The portion of the Yucatan Channel flow that does not enter the Gulf becomes, according to this model, the Loop Current. It is determined by continuity, so that

$$LC_1 = (1-\beta)YC_1$$
$$LC_2 = (1-\beta)YC_2$$
$$LC_3 = 0 \qquad (5.9)$$
$$LC_4 = 0$$

Because all of the Loop Current flow below the sill depth of the Florida Straits is assumed to be diverted into the Gulf interior, LC_3 and LC_4 are zero. The Loop Current consisting of LC_1 and LC_2 does not enter the box model domain, but serves as a transport path, or conduit, to bring fluid to the box model. Fluid leaves the box model and 'rejoins' the Loop Current as it exits through the Florida Straits. In this manner continuity is maintained. The final concentration of oxygen, temperature, and salinity in the Florida Straits is not an independent variable, but simply the result of mixing so that

$$C_1^{out} \cdot FS_1 = LC_1 \cdot C_1^{in} + u_1^{out} \cdot C_1$$
$$C_2^{out} \cdot FS_2 = LC_2 \cdot C_2^{in} + u_2^{out} \cdot C_2$$
$$u_3^{out} + u_4^{out} = YC_R \qquad (5.10)$$
$$FS_1 + FS_2 = FS = YC - YC_R$$

Here FS_1 and FS_2 are the Florida Straits fluxes, which are determined by the box model, and $C_1(out)$ and $C_2(out)$ are the concentrations. The Yucatan return flow is the output of the box model between the two sill depths, 800 m at the Florida Straits, and 2000 m at the Yucatan Channel. In order to derive equation (5.10) we assumed that during the Loop Current's transit from the Yucatan Channel to the Florida Straits, the tracer concentrations did not change by reaction or through vertical mixing. This is a robust assumption given the fairly rapid transit of the Loop Current compared to the time required to flush the Gulf interior. Therefore any concentration differences actually measured at the Florida Straits are solely due to mixing of the Loop Current with water leaving the Gulf interior. The fact that the oxygen concentrations, temperature, and salinity at the Florida Straits are between that of the Yucatan Channel and the Gulf interior (see Table 5.3) is confirmation that this assumption is reasonable.

The turnover times for each box, i.e., the time for the incoming flow to completely exchange the water volume, is of interest. The turnover times are defined as follows:

$$\tau_1 = \frac{V_1}{u_1^{in} + d_{12}}$$

$$\tau_2 = \frac{V_2}{u_2^{in} + d_{23}}$$

$$\tau_3 = \frac{V_3}{u_3^{in} + d_{34}}$$

$$\tau_4 = \frac{V_4}{u_4^{in} + d_{45}}$$

$$\tau_5 = \frac{V_5}{d_{45}}$$

5.3.4   Solution Strategy

To summarize, two independent parameters, consisting of the shape of the velocity profile and the parameter β, must be specified. Values for known quantities, such as the concentrations of oxygen, salinity and temperature and the volume and sediment surface area of each box, also must be specified *a priori*. Equations 5.2 through 5.10 represent a system of linear equations that can be combined into the compact form

$$A * x = b \qquad\qquad (5.11)$$

where x is the vector containing the 15 unknowns, arranged so that the order of magnitude of each term is similar

$$x = \begin{bmatrix} u_1^{out} \\ u_2^{out} \\ u_3^{out} \\ u_4^{out} \\ u_4^{in} \\ d_{12} \\ d_{23} \\ d_{34} \\ d_{45} \\ w_{54} \\ FS_1 \\ FS_2 \\ k_w \cdot V_1 \\ k_b \cdot A_2 \\ r_p(O_2) \cdot V_1 \end{bmatrix}$$

With the exception of d_{12}, d_{23}, and d_{34}, each of the unknowns must be positive. This provides an additional constraint on the system of equations. The vector b contains the known part of each equation and A is the matrix of coefficients derived from rearranging (5.2) through (5.10). This particular system of equations is classified as overdetermined, i.e., there are ten more equations than unknowns. This means there is no value for x that exactly satisfies equation (5.11), but there is a value that minimizes the error of the model. One could reduce the number of equations to exactly match the number of unknowns, but that would require a decision as to which equations to eliminate and would significantly reduce the confidence in the answer. Consequently, all the equations are retained and a least squares solution method is used to find the values for x that minimizes the error in reproducing the vector b, i.e.,

$$\varepsilon = A * x - b$$

where the error, ε, is minimized with respect to the parameter β, the fraction of Yucatan Channel inflow diverted into the Gulf interior. Therefore, varying the parameter β until the error becomes as small as possible further reduces the error to a global minimum. This is the procedure used in this study to establish a set of values for such parameters as transport and rate constants.

5.4   Baseline Model Run

To determine the values for the transports, vertical fluxes, reaction rates, and net surface productivity that best reproduces the known oxygen concentrations, a series of model runs were completed. These runs were made to establish the baseline, i.e., the values for the transports, vertical fluxes, reaction rates, and photosynthetic productions that best minimizes the error. For each velocity profile, a value for the parameter β was found over the range from zero to one that resulted in a global minimum for the error. Using this "minimum β", the resulting values for the unknowns are shown in Table 5.4 for each of the five velocity profiles.

The most striking result from the baseline model run is that many of the box model unknowns are independent of the velocity profile. Consider that:

1. Approximately 26 – 30% of the assumed 30 Sv Yucatan Channel flow is diverted into the interior; the remainder becoming the Loop Current. Ochoa et al. (2001) found that the net average transport through the Yucatan Channel was close to 25 Sv, consisting of an inflow of 33 Sv and a return flow of 8 Sv. Whether by coincidence or not, the 8 Sv return flow is 25% of the 33 Sv inflow and surprisingly close to the 26 – 30% value.
2. The flux out of the Gulf interior, between the surface and 200 m, is in the range from 4 to 5 Sv. This flux combines with the Loop Current to become the surface part of the Florida Straits flux.
3. The flux out of the Gulf interior, between 200 and 800 m, is 5 Sv. This flux combines with the Loop Current to become the deep part of the Florida Straits flux.
4. The vertical flux crossing the 200-m depth is towards the surface and is ~1 Sv.
5. The vertical flux crossing the 800-m depth is towards the surface and at 2.1 to 2.3 Sv.
6. The residence time, or the overturning age, of the volume between the surface and 200 m is 15 to 30 yrs.
7. The residence time, or the overturning age, of the volume between the 200 and 800 m is 25 to 30 yrs.
8. The first order rate constant parameterizing the oxidation of particulate and dissolved organic carbon is 0.065 yr^{-1}, determined from the $k_w V_l$ value after dividing by the volume of box 1. The rate constant simply means that if the Gulf were cutoff from all sources of oxygen, then it would take ~10 years for the oxygen concentration to decrease by one-half as the result of reactive loss. Given the surface oxygen concentration of 3.87 mL L^{-1}, the estimated oxygen consumption rate in the surface layer is therefore 5.3 μmol kg^{-1} yr^{-1}. This value is an order of magnitude smaller than the estimated rate of consumption estimated by Riley (1951) as presented in Table 4.3. However it is quite consistent with

the results of Jenkins (1982), who used tritium-helium3 dating to get a rate of 5 μmole/kg/yr. Subsequent studies all come up with the same number usually in the range of 1 to 5 μmole/kg/yr at 900 – 1000 m depth.

9. Based on the first order rate constant parameterizing the loss of oxygen at the sediment interface, an equivalent rate constant for reactive loss in the waters of the deep Gulf is ~0.001 yr^{-1}. Given the oxygen concentration of 5.04 mL L^{-1} for the deep Gulf, the estimated oxygen consumption rate in the deep Gulf is 0.1 μmol kg^{-1} yr^{-1}. This corresponds to a residence time on the order of 50 years. Fiadiero and Craig (1978) calculated an oxygen utilization rate below 1000 m of 2 μL/L/yr, or 0.09 μmole/kg/yr, which is consistent with the value determined here.

10. The net primary productivity and atmospheric exchange for the upper 200 m of the entire Gulf is between 1.6 to 1.9 x 10^{10} mL s^{-1}. This corresponds to 2.2 to 2.6 mol m^{-2} yr^{-1}. These values are consistent with the fluxes reported by Bopp et al. (2002) for various regions of the world's oceans and in particular the latitude associated with the Gulf of Mexico.

Again we reiterate, these results hold regardless of the assumed velocity profile. This is evidence of the self-consistency of the model.

Table 5.4

Box Model Output from Baseline Runs

Description	Units	n=1	n=2	n=3	n=4	Sheinbaum et al. (2002)
β_{min}		0.29	0.29	0.29	0.26	0.30
u_1^{out}	Sv	2.22	3.16	4.01	4.29	4.82
u_2^{out}	Sv	5.36	5.95	5.92	5.05	5.01
u_3^{out}	Sv	5.29	2.70	1.05	0.18	1.05
u_4^{out}	Sv	3.51	1.49	048	0	1.00
u_4^{in}	Sv	1.26	0.32	0.08	0.02	0.76
d_{12}	Sv	0.58	0.83	1.04	1.12	1.25
d_{23}	Sv	2.10	2.29	2.36	2.15	2.12
d_{34}	Sv	-1.62	-1.02	-0.36	0.03	0.14
d_{45}, w_{54}	Sv	0.63	0.16	0.04	0.01	0.38
FS_1	Sv	6.36	9.06	11.49	13.59	13.40
FS_2	Sv	15.10	17.06	17.33	16.65	14.78
YC_R	Sv	8.80	4.19	1.53	0.18	2.05
$k_w V_1$	Sv	0.402	0.457	0.470	0.427	0.423
$k_b A_2$	Sv	0.036	0.009	0.002	0.0006	0.022
$r_p(O_2)V_1$	mL·L^{-1} Sv	1.585	1.808	1.865	1.707	1.684
τ_1	years	29	20	16	15	13
τ_2	years	30	26	26	29	29
τ_3	years	20	30	47	75	56
τ_4	years	33	74	228	4300	109
τ_5	years	350	1390	5,535	22,560	570
ε, error		4.916	3.495	2.250	1.950	1.578

There are several other items to note from Table 5.4. First, the smallest total error is obtained for the most realistic velocity profile, the Sheinbaum profile. Since this profile was determined graphically from a figure, actual flux values would lead to better estimates of the unknowns. Secondly, the box model determined a Yucatan return flux of 2.05 Sv for the Sheinbaum velocity profile, which is surprisingly close to the 2 Sv found by Ochoa et al. 2001. Third, the flux into the bottom most box, w_{45}, while less than 0.6 Sv, still varies are over a wide range depending on the assumed velocity profile. Recent model studies by Welsh and Inoue (2000) support an extremely rapid turnover (~10 yr) of the bottommost layer, with rings as the mechanism for driving the deep circulation. On the other hand, Schink (private communication) used radiochemistry to estimate the turnover to be on the order of 70 yr. None of the box model turnover times (or residence times) for box 5 shown in Table 5.4 are close to these independent values. This implies that the vertical flux, w_{45}, into box 5 must be much larger than even the n=1 velocity profile allows. In order to obtain a turnover time of 100 years, the flux through box 5 would have to be 2.2 Sv. Assuming that 1.75 Sv came from the bottom 500 m of a Loop Current eddy over a two-year spin down period and at any one time 3 warm-core eddies were in the Gulf, then the initial size of each eddy would have to be ~300 km in diameter. This is a probable eddy diameter, but a turnover time of ten years calls for a significantly greater flux and either many more warm-core eddies or larger diameter eddies. An improvement to this box model might incorporate a generic eddy that injected fluid below the sill depth, contained Loop Current water properties, and would allow for significant vertical fluxes.

5.5 Scenarios

Next we examine the dependence of dissolved oxygen concentrations on the Loop Current transport, the rate of primary productivity, the rate of remineralization, and the rate of sediment reaction. The values in Table 5.4 serve as the baseline for testing scenarios, such as how the oxygen concentrations might change given a change in photosynthesis. The complete set of equations, described by (5.2) through (5.6), can be greatly simplified to only track the transport, production, and reactive loss of oxygen since the other values are now known. This simplified equation set is:

$Box \cdot 1$

$$0 = u_1^{in} O_1^{in} - u_1^{out} O_1 + d_{12}\{O_1, O_2\} - k_w V_1 O_1 + V_1 r_P(O_1)$$

$Box \cdot 2$

$$0 = u_2^{in} O_2^{in} - u_2^{out} O_2 + d_{23}\{O_2, O_3\} - d_{12}\{O_1, O_2\} - V_2 k_w O_2 - \alpha_2 A_2 k_b O_2$$

$Box \cdot 3$

$$0 = u_3^{in} O_3^{in} - u_3^{out} O_3 + d_{34}\{O_3, O_4\} - d_{23}\{O_2, O_3\} - V_3 k_w O_3 - \alpha_3 A_3 k_b O_3$$ (5.12)

$Box \cdot 4$

$$0 = u_4^{in} O_4^{in} - u_4^{out} O_4 - d_{34}\{O_3, O_4\} + d_{45} O_5 - \alpha_4 A_4 k_b O_4$$

$Box \cdot 5$

$$0 = w_{45} O_4^{in} - d_{45} O_5 - \alpha_5 A_5 k_b O_5$$

The transports, rate constants, and oxygen production in the surface layer are found in Table 5.4. Both the water column and sediment-water interface reactions are given with the understanding that only one of the two can be included.

Discussion of Model Runs

First examined was how the errors in the model affect the estimated oxygen concentrations in each of the boxes. As noted above, the unknowns will not completely satisfy the governing equations. If the transports, rate constants, and oxygen production in the surface layer in Table 5.4 are input to equation (5.12), then the oxygen concentrations in each box can be determined. The results are shown in Table 5.5 for each of the five velocity profiles.

Table 5.5

Model Estimated Dissolved Oxygen Concentrations by Box

Dissolved Oxygen (mL·L^{-1})	Known	n=1	n=2	n=3	n=4	Sheinbaum et al. (2002)
Box 1	3.87	3.86	3.86	3.86	3.86	3.86
Box 2	3.02	2.99	2.99	2.99	2.99	3.00
Box 3	4.42	4.31	4.31	4.31	4.32	4.34
Box 4	4.95	4.66	4.52	4.48	4.95	4.95
Box 5	5.04	5.03	5.03	5.02	5.04	5.03

If there were no errors in the box model, then the predicted oxygen concentrations for each of the velocity profiles would exactly match the expected values. As it is, the match is quite good, except for box 3. The model has difficulty in getting the oxygen concentration in box 3 high enough. In fact, the concentration of 4.31 mL·L^{-1} is that of the incoming Yucatan oxygen concentration. The only way to raise the value is to bring fluid up from the lower box, box 4 where the oxygen concentration is 4.95. But the temperature distribution is in the opposite direction and forces fluid down from box 3 to 4. This is the case for the n = 1, 2, and 3 velocity profiles, where the d_{34} vertical flux is negative. This mismatch might be corrected with a different distribution of boxes in the vertical. It also points out how averaging the vertical distribution of a tracer concentration over a large depth can lead to problems.

To examine a series of scenarios and still recognize the inherent errors in this box model, the concentrations were corrected by the error implied in Table 5.5.

For the first scenario, the role transport alone plays was examined. There is no production of oxygen in the surface layer, there are no biogeochemical processes consuming oxygen, and the transports are given by the values in Table 5.4. The results of solving equation (5.12) given these constraints are shown in Table 5.6 for each of the velocity profiles. They indicate that transport alone is sufficient to maintain the oxygen in box 3 (800 to 1500 m). This was a design constraint found by running the model a sufficient number of times to determine reasonable results that could not be obtained if any reaction was allowed in this box (see Section 5.2.3c). Above 800 m significant reactive losses must act to reduce the oxygen concentrations and below 1500 minor reactive losses must exist.

107

Table 5.6

Role of Transport in Establishing Dissolved Oxygen Concentrations

Dissolved Oxygen (mL·L^{-1})	Known	n=1	n=2	n=3	n=4	Sheinbaum et al. (2002)
Box 1	3.87	4.00	4.00	4.00	4.00	4.00
Box 2	3.02	3.58	3.58	3.58	3.59	3.58
Box 3	4.42	4.42	4.43	4.43	4.43	4.43
Box 4	4.95	5.04	5.00	4.99	5.12	5.12
Box 5	5.04	5.13	5.13	5.14	5.13	5.13

The role that photosynthesis plays was examined next. Here, there is no photosynthetic production or atmospheric exchange of oxygen in the surface layer, but there is reactive loss. The rate constants and transports are given in Table 5.4. The results of solving equation (5.12) given these constraints are shown in Table 5.7. The oxygen concentration in the surface layer is reduced by reactive loss from the 4.00 mL·L^{-1} that transport mixes into the surface layer to a value that is lower than the measured concentration of 3.87 mL·L^{-1}. The remaining boxes are unaffected because the vertical flux into the surface layer is positive upwards for every velocity profile (see d_{12} in Table 5.4). Therefore, a change in the surface layer cannot be communicated downward and the surface layer is in essence isolated from the underlying boxes as long as the net vertical flux is upwelling. It may well be that there are local regions where the flux is downwelling, but the nature of the box model essentially assumes a global net flux.

The contribution of the sediment interface reaction was next examined. For this scenario, there is no sediment reaction. The rate constant for the water column reaction, the net production of oxygen in the surface layer, and the transports are given in Table 5.4. The results of solving equation (5.12) given these constraints are shown in Table 5.8. The only significant difference is seen in the bottom two boxes where the sediment reaction occurs. A slight increase in the concentrations is seen.

A crude estimation of what would happen if there was a doubling of carbon in the water column can be made by doubling the rate constant for the water column reaction in the top two boxes and including it in the bottom three boxes (Table 5.9). This assumes that there would be significant labile particulate carbon left by the time it settled below the second box. As we have noted previously, this model cannot directly track the fate of hydrocarbon in the water column unless carbon is tracked simultaneously with oxygen. Nevertheless, this scenario shows a dramatic decrease in the oxygen concentrations, especially at depth. This reinforces our findings that the water column reaction actually contributes very little below 800 m in depth. It also infers that an introduction of carbon at depth has the *potential* to significantly decrease the oxygen content if the system is closed.

Table 5.7

Scenario of No Oxygen Production in the Surface Layer

Dissolved Oxygen (mL·L^{-1})	Known	n=1	n=2	n=3	n=4	Sheinbaum et al. (2002)
Box 1	3.87	3.26	3.37	3.45	3.50	3.54
Box 2	3.02	3.02	3.02	3.02	3.02	3.02
Box 3	4.42	4.42	4.42	4.42	4.42	4.42
Box 4	4.95	4.95	4.95	4.95	4.95	4.95
Box 5	5.04	5.05	5.05	5.06	5.04	5.04

Table 5.8

Role of Sediment-water Reactions in Maintaining Dissolved Oxygen Concentrations

Dissolved Oxygen (mL·L^{-1})	Known	n=1	n=2	n=3	n=4	Sheinbaum et al. (2002)
Box 1	3.87	3.87	3.87	3.87	3.87	3.87
Box 2	3.02	3.02	3.02	3.02	3.02	3.02
Box 3	4.42	4.42	4.42	4.43	4.43	4.43
Box 4	4.95	5.04	5.00	4.99	5.12	5.12
Box 5	5.04	5.13	5.13	5.14	5.13	5.13

Table 5.9

Role of Carbon Doubling in Maintaining Dissolved Oxygen Concentrations

Dissolved Oxygen (mL·L^{-1})	Known	n=1	n=2	n=3	n=4	Sheinbaum et al. (2002)
Box 1	3.87	3.25	3.31	3.36	3.37	3.42
Box 2	3.02	2.50	2.43	2.34	2.23	2.30
Box 3	4.42	3.95	3.68	3.33	2.95	3.18
Box 4	4.95	3.58	2.82	1.74	0.14	2.37
Box 5	5.04	1.62	0.49	0.14	0.04	1.09

Finally, an estimate was made of the decay half-life of the oxygen. Here the questions asked are: "What would happen if there was no more transport of oxygen from the Yucatan into the Gulf and no more photosynthesis?" and "How long would it take for the oxygen to decay?" For the water column reaction the decay of oxygen is given by

$$V\frac{dC}{dt} = -k_w VC$$

which has the solution

$$C(t) = C(t_0)e^{-k_w t}$$

The concentration drops to half of its initial value at the half-life, given as

$$t_{1/2} = \frac{\ln 2}{k_w}$$

Similarly, the half-life for the sediment contribution is

$$t_{1/2} = \frac{\ln 2}{k_b \dfrac{A}{V}}$$

The results for each of the velocity profiles are shown in the Table 5.10. The large variation in the sediment half-life is not surprising given the range of mass transfer coefficients shown in Table 5.4 and the large uncertainty in the vertical flux. The time necessary for the oxygen concentration in the water column to decrease to half of its initial value ranges from 9.5 to 11.2 years, depending on the velocity profile. This range is entirely due to the range in rate constants reported in Table 5.4.

Table 5.10

Time Necessary for Dissolved Oxygen Concentration to Decrease by 50% If There Is No Transport

Half-life (years)	n=1	n=2	n=3	n=4	Sheinbaum et al. (2002)
Water column	11.2	9.8	9.5	10.4	10.6
Sediment	123	500	2,025	7,836	200

5.6   Conclusions

Despite its apparent complexity, this is a very simple box model. However, while it is a zero-order approach, the results should not be summarily dismissed as zero-order. By reason of the model's design philosophy, the results provide a reasonable bound to the problem of how oxygen is maintained in the deep Gulf. The model was deliberately developed with the least amount of external information possible. In this way we were assured that the principles of physics and biology would constrain and guide the box model development, rather than some assumed knowledge about the system. In every case, as the box model evolved to what is reported here, the model had to be adjusted in ways that were then found to be realistic. For example, the water column reaction had to be severely attenuated below 800 m in order to obtain reasonable results. This is representative of the rapid consumption of labile carbon in the upper water column.

The results from the box model can be roughly separated into two parts, findings that are most likely independent of the specific details of the box model, and findings that may not be independent. Regarding independent findings, the box model indicates to us that the oxygen profiles in the Gulf are most likely determined by a unique combination of a) the transport process associated with circulation and mixing and b) the reactive processes associated with biogeochemical cycles. It is unlikely that there are a multitude of oceanographic and biogeochemical states that could lead to the observed oxygen profiles. From a mathematical perspective this means this is a boundary value problem. From a physical perspective it means that extra-basin effects, in particular the transport through the Yucatan Channel, have a significant role in controlling the oxygen profiles in the Gulf. This is seen in how the shape of the assumed velocity profile is critical to determining how well the box model reproduces the observed oxygen concentrations. Obtaining accurate flux information in the Yucatan, and for that matter, through the Florida Straits, should be a priority if the box model is to be improved.

The results of the box model suggest that ventilation of the deep Gulf waters below the 2000-m sill depth of the Yucatan Channel is key to maintaining the dissolved oxygen distributions. First, there is the matter of the lower oxygen concentration in the Gulf above 2000 m. The source of the higher concentration oxygen in the deep Gulf waters must come directly from a water mass that contains properties similar to the lowest level of the Yucatan Channel. Independent estimates of the residence time in the deep Gulf that range from 10 to 70 years indicate that the flux below the 2000 m depth has to be much larger than the box model estimates of 0.01 to 0.6 Sv. In its present design, the vertical flux across the 2000-m depth is constrained by the modeled flow at depth through the Yucatan Channel. This constraint could be eased by allowing for an 'autonomous' source of high oxygen concentration water. The vertical flux could be determined by the box model, but, as the model is presently formulated, to do so would presume that the sediment-mediated rate constant controls the flux. Considering equation (5.12) for box 5, the vertical flux would be given by the following:

$$w_{45} = \frac{\alpha_5 A_5 k_b O_5}{O_4^{in} - O_5}$$

Here we see that the flux is entirely dependent on the value for the sediment rate constant. Since this is the only equation defining the flux, it is unlikely to provide a statistically reliable answer. Because the mechanisms for transport of deep water into the Gulf interior and for vertical mixing below the sill depth are unknown, the necessary flux cannot be resolved by the box model. Determining the flux will require a much more sophisticated numerical model than a box model.

There are three results from the box model that are quite enlightening, though they may not be independent of the box model design. First, approximately 30% of the Yucatan Channel flow is diverted into the interior; the remainder outflowing as the Loop Current. Secondly, the first order rate constant parameterizing the oxidation of particulate and dissolved organic carbon is 0.065 yr^{-1}. Given the surface oxygen concentration of 3.87 mL·L^{-1}, the estimated oxygen consumption rate is therefore 5.3 μmol kg^{-1} yr^{-1}. This value is comparable to the estimated rate of consumption found by Riley (1951); seen Table 4.3. Third, the net primary productivity and atmospheric exchange for the upper 200 m of the entire Gulf is between 2.2 to 2.6 mol m^{-2} yr^{-1}.

6 DISCUSSION, CONCLUSIONS, AND RECOMMENDATION

This section provides a brief discussion of issues associated with dissolved oxygen in the Gulf of Mexico (Section 6.1), conclusions from the study (Section 6.2), identification of significant gaps in data and knowledge (Section 6.3), and a recommendation for further study (Section 6.4).

6.1   Discussion

Seven questions were developed to evaluate what this study has revealed about the types and rates of processes occurring in the deep Gulf of Mexico that affect the levels of dissolved oxygen in the deep water and about the balance that maintains this level. The knowledge that was obtained in this study for each question also was examined.

What major processes affect the levels of oxygen in the deepwater Gulf of Mexico?
The major processes that affect the dissolved oxygen concentrations in the Gulf of Mexico have been known for over 50 years. This study's findings are consistent with earlier studies. The major sources are water mass transport into the basin from the Caribbean Sea (source waters for the deep Gulf), the atmospheric-oceanic exchange of oxygen at the air-sea interface (ocean generally in equilibrium with the atmosphere for the environmental temperature and salinity conditions), and primary productivity and its consequent photosynthesis input of oxygen (surface phenomenon with downward advection only within the waters mixed by atmospheric processes). Aside from transport of water masses out of the basin, the major sinks are removal by respiration of organic matter in both the water column and the sediments. Other processes, such as chemical removal or organic enrichments from oil and gas seeps, are minor or only potentially of local importance.

The distribution of the data sets, particularly within the deep waters of the Gulf, is relatively sparse in time and space. However, the relative uniformity of the temperature and salinity in the deep waters below about 1500 m and the slow rates of oxygen consumption that occur at those depths suggest that the distribution is adequate to resolve the horizontal and vertical oxygen distributions in those deep waters.

What is the relative importance of these processes? What are the timeframes for these processes – hours, days, years, decades, centuries?
The relative importance of the processes that maintain the dissolved oxygen distributions through the water column are dependent on the depth. In essence, the Gulf can be viewed as a two layer system when it comes to considering major processes of dissolved oxygen replenishment. These are the upper layer, which is approximately the upper 200 m, and the lower layer, which extends from approximately 200 m to the bottom. The major process of oxygen replenishment in the upper layer is through air-sea exchanges of oxygen and mixing by wind and wave action. This process is continually present and maintains the upper waters at or near equilibrium concentrations. Input of oxygen by photosynthesis can be important in this layer, but it is local in nature. It can result in supersaturation of waters, particularly below the actual air-sea interface.

The only process for renewal of dissolved oxygen to the deep water is advection into the Gulf interior of inflowing source waters. There is no deep water ventilation by water mass formation and thermohaline circulation. The processes that advect the inflowing waters into the interior are not well understood. The transport into the basin at the Yucatan Channel is balanced by the outflow at the Florida Straits and recirculation out the Yucatan Channel, and although the transport out the Florida Straits is reasonably understood, the recirculation is not. Further, the transport mechanisms that move the well-oxygenated deep waters from the Yucatan Channel into the Gulf interior, particularly into the western Gulf, are not understood.

The major sink of dissolved oxygen throughout the water column is the oxidation of organic material. This process is not unique to the Gulf, nor does the Gulf have unique aspects, except for the localized hydrocarbon seeps and anoxic brine pools. These seeps and pools do not affect the dissolved oxygen levels in the water column except within a few centimeters above them. This oxidation of organic material also operates, in essence, in two different layers. The first layer again is generally in the upper 200 m, which is where, in the world's ocean, approximately 90% of the organic matter from photosynthesis is oxidized (Millero 1996; Riley 1951). Below this layer and in particular below the oxygen minimum zone, which can occur as deep as 700 m in the Gulf, the oxidation rate is assumed to decrease exponentially. The water column respiration, however, is not well measured; no direct measurements in the Gulf were found. At the sediment-water interface, the oxygen consumption processes of the sediments can be of local importance in drawing down the oxygen concentrations within a few cm of the sediment surface. In a totally closed ocean basin, this consumption over time could result in anoxic conditions in substantial portions of the deep layer, such as seen in the Black Sea. However, the Gulf of Mexico basin is not such a closed system, and oxygen removed by these processes is replenished by the advection into the interior of the oxygen-rich water masses.

The temporal rates of the processes involved in maintaining the oxygen distributions are not well known. However, given the continual input of natural hydrocarbons and the maintenance of the well-oxygenated deep waters, it is likely that the intra-basin effects of consumption are in balance with the extra-basin processes that transport into the deep Gulf the oxygen-rich water masses that are formed at high latitudes. Flushing times of the deep Gulf are uncertain, in part because the processes of advection into this region are not well understood.

What are the relevant spatial and temporal scales of these processes?
The station location maps show that most areas within the deepwater Gulf have been sampled at least once. Although more stations were taken in summer (~30-40%), sufficient numbers were taken in the other seasons (15-26% each) to reduce the seasonal bias in sampling. However, the existing data base remains lacking in both the spatial and temporal coverage necessary to address questions of scales of the processes. A crucial factor that is missing is a good understanding of the processes of circulation in the deep waters of the Gulf. The circulation in the upper 1000 m is reasonably well-defined, but the circulation and its variability in lower waters are poorly measured and, so, are not well understood. Most of the sampling for dissolved oxygen has been on cruises with fairly wide spacing between stations, and very few cruises sampled significant portions of the Gulf quasi-synoptically. Vertically, the sampling is sparse below 1500 m, but the relative uniformity of the concentrations suggests that the sampling done has been adequate to reveal the major distributions.

Because the dissolved oxygen system in the Gulf of Mexico operates very much as it does in the world's ocean, it is unlikely that major processes or potentially catastrophic events have been missed by the sampling. It is likely that events or distributions that occur over very short timeframes or small spatial scales may have not been sampled; but their importance also is likely to be small to negligible for the broad oxygen distributions. However, they may be important for local considerations, such as timeframes of several years or less over a region of several kilometers or less. For example, hurricanes can have effects on the distribution of oxygen, both through the injection of bubbles into the water column and from storm-induced upwelling (e.g., see Ichiye 1972). As another example, the brine pools in the Gulf are anoxic, however, their major effects appear to occur on small scales (cm) in their immediate vicinity, and the replenishment of dissolved oxygen through the general circulation that transports oxygen rich waters into these areas appears to be sufficient to prevent the anoxic conditions from growing in extent. However, the available data base does not allow a clear determination of the rates of replenishment or assessment of the possibility that one time events set the distribution and then the profiles begin to degrade.

114

Do we expect oil and gas related discharges to significantly alter oxygen levels in the deep Gulf?
The oil and gas remaining proven and unproven reserves for the U.S. Gulf of Mexico have been estimated by MMS at 5,225 million barrels of oil and 28,699 billion cubic feet of gas (Crawford et al. 2003). Using the barrels of oil equivalent (BOE) conversion factor of 5,620 cubic feet gas/BOE, yields a BOE of 5,107 million barrels from the gas. Thus, the total quantity of hydrocarbons in these reserves is approximately 10 billion barrels. Complete oxidation of these oil and gas reserves would consume 0.76% of the standing oxygen supply. By comparison the annual discharges from extraction (drill fluids including synthetic-based mud, drill cuttings, produced waters, accidental spills from platforms, etc.) would consume 0.000001% of the standing oxygen supply. These effects are not significant alterations of the oxygen supply, unless the Gulf becomes a closed system that is not ventilated by the inflowing oxygen-rich water masses through the Yucatan Channel.

There can be local effects, such as near platforms where cuttings might be discharged, smothering the benthic communities, and resulting in draw down of oxygen just above the sediments (NRC 2003; Neff et al. 2000). However, the database is insufficient to evaluate local events and their effect on oxygen distributions in the immediate vicinity. The database does not include the sampling to give the details of dissolved oxygen distributions in deep water. The nature of the hydrocarbon material, the lability of the material, its chemistry and properties within the temperature and salinity structure of the deep water, the location of its discharge in the water column, its likely movement (rising to the surface, staying near the sea floor or at a density level in the water column if formed into dense enough emulsions with water or sediments, advecting to distant locations through circulation processes)--all these factors and more influence the extent of the local impact of such material on dissolved oxygen concentrations. The environmental conditions at such locations also would play a substantial role in controlling the magnitude of the local effects. For example, the likely impact of currents at the deep water location may be to disperse cuttings in deep water before they accumulate, or lack of light or nutrients may hinder growth of the organisms that oxidize the organic material.

Does the box model indicate any unrecognized sources or sinks for oxygen in the Gulf of Mexico? Are there regional differences?
The limitations of the data and gaps in knowledge, including a paucity of specific information on the deep circulation and transport rates, lack of availability of specific rates of consumption within the water column, and insufficient high quality oxygen data sets for all water depths, regions, and sediment-water interface, resulted in development of a first order simple box model for the whole Gulf. The model therefore was not capable of testing to determine whether there were unrecognized sources or sinks or regional differences. It also was not designed to test local conditions. Two issues that might be interesting to address with a model are (1) what the intra-basin variations might be and what causes them and (2) determination of whether micro-scale environments, such as the brine pools, might drive mesoscale features. However, limitations of the data and gaps in knowledge and, consequently, this simple box model are not adequate to address these issues in detail.

Do oxygen distributions indicate any areas unusually sensitive or resilient to oxygen depletion and what are the characteristics that make these areas unique?
Evaluation of the database suggests that the deep waters in the southwest Gulf may contain older water that is slightly less well-oxygenated than the southeast or northern Gulf. However, the data are not of sufficiently high quality or quantity to determine this definitively. If real, this distribution makes sense because the southwest Gulf likely is ventilated by transport more slowly than are the other areas. The deep circulation is not well known and there are no definitive studies on the age of the deep waters and their residence time. The available oxygen database cannot address these issues. The data base also does not reveal localized unusual areas, although other studies have found such features as anoxic brine pools and natural hydrocarbon seeps with chemosynthetic (and oxygen using) communities in local areas.

What information is missing to develop a quantitative inventory of oxygen in the basin?
The major information missing is a thorough understanding of the deep water circulation, which controls how and at what rates the well-oxygenated (~5 mL·L⁻¹) deep waters inflowing at the Yucatan Channel are transported into the Gulf interior to ventilate the waters below ~1500 m. Additionally, definition of the rates of consumption with depth are specifically needed for the Gulf, with its large hydrocarbon input through the seeps. To address local effects, detailed studies of the fates and effects of anthropogenic inputs near discharge points are needed.

6.2   Conclusions

The deep waters of the Gulf of Mexico Basin are ventilated by the circulation and mixing into the Gulf interior of the well-oxygenated water masses inflowing at the Yucatan Channel. Consequently, the Gulf of Mexico Basin is in no danger of becoming anoxic due either to natural processes or oil and gas production. Nature has been providing more oil to the basin in the form of seeps than is being discharged by humans, and yet the values of dissolved oxygen throughout the water column are stable. This is a natural experiment of geological age. Decreased oxygen levels, however, could occur on localized bases.

Sources of Dissolved Oxygen in the Deepwater Gulf of Mexico
•   The Gulf of Mexico is a semi-enclosed sea with two ports. The major inflow port is at the Yucatan Channel, which has a sill depth that is deep enough (~2000 m) to allow the transport of oxygen rich waters into the Gulf. The Loop Current brings these waters in. The major outflow port at the Florida Straits is shallow enough (~800 m) that the deeper, oxygen rich waters can mix into the interior of the Gulf, rather than flowing directly out with the Loop Current.

•   The source of dissolved oxygen to the upper waters (~100-200 m) in the Gulf of Mexico is mainly from exchanges with the atmosphere with local contributions from photosynthesis.

•   The source of dissolved oxygen to the deep waters is the transport of oxygen-rich water masses into the Gulf of Mexico from the Caribbean Sea through the Yucatan Channel. There is no water mass formation in the Gulf of Mexico to replenish the deep oxygen concentrations. So, the deep circulation of the Gulf and its associated mixing are the mechanisms that replenish the deep oxygen.

Sinks of Dissolved Oxygen
•   The major sink of oxygen in the Gulf, as elsewhere in the world's oceans, is oxidation of organic matter.

•   The oxygen minimum zone, which generally is found in the Gulf between 300-700 m, is derived from two mechanisms. First, the Tropical Atlantic Central Water with the oxygen minimum at its core brings in waters that are depleted from processes occurring outside the Gulf. The decay of organic matter that occurs in the Gulf itself augments this minimum by the same processes. However, the productivity of the Gulf is not high enough to create extreme oxygen minimum zones that are seen in other parts of the world's ocean, such as in the Arabian Sea.

Data Sets
•   Vertical sets of dissolved oxygen data, from bottle samples, were collected at approximately 4000 stations in the Gulf of Mexico.

•   These data were collected from 1922 through 2001, when this study ceased data assembly. The full water column was most likely to have been sampled on cruises during the decades of the 1950s, 1960s, and 1970s. Only the decades of the 1960s and 1970s had station coverage throughout the Gulf basin.

116

•   The southwest Gulf is less sampled than other areas.

•   Waters at or below 1500 m are not well sampled, with less than 500 stations sampling at those depths.

•   Data quality is variable, even within cruises. This likely is due to different analysis techniques used in the older data sets and differences in the analysts' visual identification of the end points of titration, as well as poor sampling and/or laboratory analysis.

•   Profiles of dissolved oxygen from sensors were obtained for approximately 1800 stations. It is not clear whether these have been properly calibrated with bottle data.

Spatial and Temporal Changes in Oxygen Levels
•   Early studies, using known high quality data sets, found that below approximately 1500 m there was "no clearly discernable horizontal variation in dissolved oxygen in these waters" and "only slight vertical oxygen gradients" (Nowlin et al. 1969).

•   Analysis of the additional data sets used in this study suggest the deep Gulf waters may have slightly different oxygen values in three different regions. However, the quality of the data sets is variable enough that this finding is not definitive. The three regions are:

- The southeast Gulf has deep oxygen concentrations most similar to those of the inflowing source waters. The close contact of the southeastern Gulf waters with the inflowing Loop Current waters likely results in more rapid replenishment of oxygen than elsewhere in the Gulf. This replenishment would be through mechanisms of direct water mass transport and the effects of mixing induced as water flows over the sill or otherwise interacts with the topography in the region.

- The deep waters of the northern Gulf have slightly less oxygen than those in the southeast, suggesting the flushing time there is longer. However, both the northeast and northwest Gulf waters have similar concentrations, and so the transport and mixing mechanisms, which carry oxygen rich waters into these regions, are the same for these two regions.

- The southwestern Gulf has deep oxygen concentrations that are less than the northern Gulf waters, suggesting a longer flushing time there.

•   Comparison of data sets from the 1970s with those from 2000/2001 indicates there has not been any discernible change in the oxygen structure in the Gulf of Mexico. This suggests that the transport mechanisms that replenish the oxygen are adequate to balance the oxygen consumption from decay of organic matter, including that from oil seeps and anthropogenic sources.

Influences of Shelf Processes on Deepwater Dissolved Oxygen Concentrations
•   Data sets available are not adequate to study the details of the effects of shelf-deep ocean exchanges on the oxygen levels.

•   No water mass is formed on the shelves. Thus, there are no shelf waters dense enough to sink down into the deep ocean of the Gulf. Any shelf-deep ocean exchanges will impact only the upper waters.

•   Transport of organic matter off the shelf, particularly in the region of the Mississippi River Delta, could have local effects on the oxygen concentrations in the deep Gulf if the material sinks to depth before decaying.

Effects of Natural Hydrocarbons on Dissolved Oxygen Concentrations
• The NRC (2003) reports an estimated 140,000 ± 60,000 tonnes of hydrocarbons are introduced into the Gulf of Mexico from natural seeps each year. This seepage begins at the sea floor and rises through the water column to the surface, so it potentially can affect the total water column. Although this natural leakage of hydrocarbons into the Gulf apparently has been occurring for millions of years, no significant large-scale perturbations in the dissolved oxygen content of the deep Gulf waters were observed by this study to result from this input. However, localized low oxygen conditions are reported to occur in the upper meter or so of the sediments and, hence, it is likely that oxygen concentrations of the waters at the sediment-water interface may be depleted.

• Chemosynthetic communities are active at the seep sites. Because they require access to oxygenated waters, their presence is additional evidence of ventilation of the deepest Gulf waters.

Effects of Oil and Gas Activities on Dissolved Oxygen Concentrations
• Knowledge of many factors is necessary to make a thorough assessment of the effects of various oil and gas activities on dissolved oxygen concentrations in the Gulf of Mexico. These factors include (1) the environmental conditions at the discharge location such as currents at all affected depths, surface and internal wave action, water temperatures, and local biological characteristics of the ecosystems that are exposed, (2) the type and rate of the discharge, and (3) the chemical nature of the material discharged. This is a highly complex problem that requires a more sophisticated approach that can be accomplished in this study.

• Effects on dissolved oxygen concentrations in the deepwater Gulf of anthropogenic discharges most likely would be most substantial at the sediment-water interface at a discharge site or the sea surface reached by the plume, not within the water column itself. Effects are expected to be local, but potentially could be severe for short periods. Further study is needed to examine such effects because they are localized and complex.

• The NRC (2003) reports anthropogenic sources of hydrocarbons are estimated at approximately 2000 tonnes from extraction (1700 from produced waters), 1600 tonnes from transportation, and 6800 tonnes from consumption. These inputs are widespread throughout the Gulf. Thus, their impact on dissolved oxygen concentrations is negligible over the large scale in the deepwater Gulf of Mexico. The localized effects of discharges, however, might be measurable and, in some cases, substantial. Most of these discharges occur at or near the sea surface, so there would be essentially no effect on the dissolved oxygen concentrations of the deep waters. An exception might be the secondary effects from any resulting increase in the quantity of decaying organic material, such as from growth of organisms that utilize the hydrocarbons, that sinks through the water column.

• Catastrophic oil spills can introduce hydrocarbons at 2-3 times the rate of the natural seeps. The effect of such a spill on the dissolved oxygen concentrations would depend on many factors, including chemical characteristics of the spilled material, location in the water column and nature of the spill, residence time in the water column of labile components of the hydrocarbon as the spilled material rises toward the sea surface, and environmental conditions at the spill site, including sites to which the material may be transported. In the large scale, the impact is expected to be minimal, although, again, local effects on oxygen levels could be substantial.

Modeling Results
The results of the simple box model developed in this study provided insights into the importance of transports, reactive losses, and primary productivity. Transport is the primary process controlling the oxygen profiles, followed in importance by respiration of organic material in the water column. The oxidation of organic matter in the sediment plays a minor role. Specifically:

•   The oxygen profiles in the Gulf likely are determined by a unique combination of the transport process associated with circulation and mixing and the reactive processes associated with biogeochemical cycles. It is unlikely that a multitude of oceanographic and biogeochemical states could describe the observed oxygen profiles. This implies that a model, if properly designed and constrained, is mathematically capable of determining a unique answer.

•   The transport through the Yucatan Channel plays the most significant role in controlling the oxygen profiles in the deep Gulf. The model suggests that 30% of the inflow is diverted into the Gulf. However, the modeled Yucatan Channel flow at depth is not sufficient itself to maintain the oxygen profiles in the deepwater Gulf. Because the mechanisms for transport of deep waters into the Gulf interior and for vertical mixing in waters below the sill depth are unknown, the necessary flux cannot be resolved by the box model.

•   The oxidation of carbon in the water column plays a secondary, but important, role to that of transport in maintaining oxygen profiles in the upper 800 m of the Gulf. It was not found to be necessary below 800 m.

•   The model is incapable of simultaneously determining, within a box, oxidation rates in the water column and oxidation rates at the sediment interface. Some judgment is necessary to choose which process is to be modeled within each box.

•   The oxidation at the sediment-water interface plays a minor, but necessary, role in maintaining the oxygen below 1500 m. It was not found to be necessary above 1500 m. By design, it has no effect above 200 m.

•   The net production of oxygen in the surface layer is consistent with the rates reported for the latitude associated with the Gulf of Mexico.

•   The oxygen content in the Gulf would decrease to half of its present value in 10 years if the Gulf was isolated from all sources of dissolved oxygen, be it transport, primary productivity, or atmospheric exchange.

6.3   Knowledge and Data Gaps

This study has revealed a number of gaps in data and knowledge regarding dissolved oxygen in the deepwater Gulf of Mexico that are of scientific interest. These include

1.   What is the rate of consumption of dissolved oxygen throughout the water column, particularly below the mixed layer and also in regions with natural hydrocarbon seeps?

2.   What is the age of the water masses, below the mixed layer, in the various regions of the Gulf of Mexico? In particular, what are the ages of the inflowing water masses at the Yucatan Channel as compared to those in the southeast, northeast, northwest, and southwest quadrants of the Gulf interior? What is the age of the outflowing Florida Strait waters?

3.   What is the circulation below 1500 m that results in the replenishment of the deep dissolved oxygen throughout the Gulf? What mechanisms are there for horizontal advection in the deep waters of the Gulf? What are the vertical fluxes involved and what mechanisms cause them? How do deep waters from the Yucatan Channel mix into the interior of the Gulf? What mechanisms operate in the four quadrants of the Gulf and how do they compare?

4.   What is the Gulf-wide horizontal structure of dissolved oxygen distributions below 1500 m and in the oxygen minimum zone?

119

5.   How will dissolved oxygen concentrations in the Gulf of Mexico change through time with global warming?

6.   What is the carbon cycle in the deepwater Gulf of Mexico and how is it maintained?

Each of these can be addressed by a combination of (1) high quality field measurements of dissolved oxygen, currents, and other parameters specific to the particular question and (2) detailed modeling studies using coupled physical-biogeochemical, primitive equation models. Some of these issues would require repeated measurements every 5 years or so to monitor whether the system is changing.

The gap of greatest application to the goals and responsibilities of the Minerals Management Service, however, concerns understanding the details of the effects of hydrocarbons on dissolved oxygen concentrations *in specific local situations*. This study has shown that, on large scales, the dissolved oxygen concentrations have not changed over the 80 years during which data have been collected. This is despite increasing anthropogenic inputs of hydrocarbons to the environment. However, the data sets available are not adequate to determine whether local changes may have occurred or to determine the local effects of specific anthropogenic inputs. Furthermore, the complex interactions between the physical, chemical, biological, and geological factors, associated with both the anthropogenic input itself and the environmental conditions, that determine whether specific anthropogenic inputs will have substantial local effects on the dissolved oxygen content of the local oceanic ecosystem, are not well understood or modeled.

6.4   Recommendation

To determine whether anthropogenic activities are locally affecting dissolved oxygen levels will require local monitoring of dissolved oxygen. A study designed to assess the nature and extent of the effects on dissolved oxygen concentrations of discharges at selected sites of oil and gas exploration and production operations in the deepwater Gulf, or perhaps at natural seeps to provide a natural experiment to determine important parameters for later modeling studies, should be undertaken. Monitoring of discharges should include measurements to determine the fate of drill cuttings, drilling fluids, produced waters, and similar discharges. Local environmental conditions, such as currents, waves, temperature, salinity, nutrients, carbon, particulate matter, and other chemical properties should be monitored at the selected locations to allow determination of the importance of the various complex of factors involved and to provide information that would allow development of an oil spill model that could assess possible effects on dissolved oxygen of a hypothetical subsurface blowout. Only if substantive effects are shown to occur should detailed modeling, process, or other studies be undertaken.

7 LITERATURE CITED

Aminot, A. 1979. Anomalies du système hydrobiologique côtier après l'échouage de L'AMOCO CADIZ. Considérations qualitatives et quantitatives sur la biodégradation in situ des hydrocarbures. In Amoco Cadiz: Consequences of Accidental Hydrocarbon Pollution. CNEXO. pp. 223-242.

Anderson, L. A. 1995. On the Hydrogen and Oxygen Content of Marine Phytoplankton. Deep-Sea Res. I, 42, 1675-1680.

Anderson, L. A. and Sarmiento, J. L. 1994. Redfield ratios of remineralization determined by nutrient data analysis. Global Biogeochemical Cycles, 8, 65-80.

Anderson, L. G. P. O. J. Hall, A. Iverfeldt, M. M. Rutgers Van De Loeff, B. Sundby, and S. T. G. Westerlund. 1986. Benthic respiration measured by total carbonate production. Limnology and Oceanography, 31, 319-329.

Austin, G. B., Jr. 1955. Some recent oceanographic surveys of the Gulf of Mexico. Transactions, AGU, 36 (5), 885-892.

Avanti Corporation. 1997a. Biodegradability of synthetic-based drilling fluids. Contract Report 68-C5-0035. Environmental Protection Agency, Office of Science and Technology, Washington, D.C. 20 pp.

Avanti Corporation. 1997b. Seabed survey review and summary. Contract Report 68-C5-0035. Environmental Protection Agency, Office of Science and Technology, Washington, D.C. 50 pp.

Baringer, M. O., and J. C. Larsen. 2001. Sixteen years of Florida Current transport at 27°N. Geophys. Res. Letters, 28 (16), 3179-3182.

Berberian, G.A., and Y.A. Cantillo. 1999. Oceanographic conditions in the Gulf of Mexico and Straits of Florida, Fall 1976. NOAA Data Report OAR AOML-36. Atlantic Oceanographic an Meteorological Laboratory, Miami, FL.

Biggs, D. C., ed. 1989. Hydrographic data from the Texas continental shelf and the NW continental slope of the Gulf of Mexico: Texas Institutions Gulf Ecosystem Research (TIGER) cruise 89G-15. Issued 19 December 1989. Tech. Rpt. 89-05-T. TAMU Oceanography, College Station, TX. 163 pp.

Biggs, D. C., ed. 1990. Hydrographic data from the Texas and Louisiana continental shelf of the NW continental slope of the Gulf of Mexico: Texas Institutions Gulf Ecosystem Research (TIGER) cruise 90G-10. Issued 2 November 1990. Tech. Rpt. 90-04-T. TAMU Oceanography, College Station, TX. 353 pp.

Biggs, D. C., ed. 1991a. Gulf of Mexico hydrographic data: XBT, CTD, and bottle data from R/V Gyre cruises 90G-14 and 90G-15. Issued 17 January 1991. Tech. Rpt. 91-01-T. TAMU Oceanography, College Station, TX. 143 pp.

Biggs, D. C., ed. 1991b. Hydrographic data from the Texas continental shelf and the NW continental slope of the Gulf of Mexico: cruise 91G-02 of the Texas Institutions Gulf Ecosystem Research (TIGER) and the analysis Multidisciplinario de Investigaciones en Golfo Occidentale (AMIGO), a cooperative initiative between Texas A&M University and the secretaria de Marina de Mexico. Issued 10 May 1991. Tech. Rpt. 91-02-T. TAMU Oceanography, College Station, TX. 329 pp.

Biggs, D. C., ed. 1991c. Gulf of Mexico hydrographic data: CTD and bottle data from R/V Gyre cruise 91G-04 and CTD data from R/V Powell cruise 91P-03. Issued 15 October 1991. Tech. Rpt. 91-05-T. TAMU Oceanography, College Station, TX. 88 pp.

Biggs, D.C., ed. 1994a. Ship-of-Opportunity hydrographic data from R/V Gyre cruises [94G-02 and 94G-03] to the NW and central Gulf of Mexico in May 1994. Issued 31 August 1994. Tech. Rpt. 94-02-T. TAMU Oceanography, College Station, TX. 226 pp.

Biggs, D.C., ed. 1994b. Hydrographic data from a July 1994 transit-of-opportunity from Key West, Florida, to Galveston, Texas: R/V Gyre cruise 94G-05, Leg Two. Issued 31 August 1994. Tech. Rpt. 94-03-T. TAMU Oceanography, College Station, TX. 303 pp.

Biggs, D.C., ed. 1994c. Hydrographic data from R/V Gyre cruise 94G-07. Issued 21 October 1994. Tech. Rpt. 94-04-T. TAMU Oceanography, College Station, TX. 202 pp.

Biggs, D.C., ed. 1995a. Hydrographic data from R/V Gyre cruise 94G-08. Issued 13 January 1995. Tech. Rpt. 95-01-T. TAMU Oceanography, College Station, TX. 296 pp.

Biggs, D.C., ed. 1995b. Cooperative studies of the circulation of the Gulf of Mexico in June 95: CTD and XBT sections through the Loop Current and Eddy Z, from R/V Gyre cruise 95G-03. Issued 14 August 1995. Tech. Rpt. 95-08-T. TAMU Oceanography, College Station, TX. 393 pp.

Biggs, D.C., and P. H. Ressler. 2001. Distribution and abundance of phytoplankton, zooplankton, ichthyoplankton, and micronekton in the deepwater Gulf of Mexico. Gulf of Mexico Science, 1, 7-29.

Boehm, P. D., and D. L. Fiest. 1982. Subsurface distributions of petroleum from an offshore well blowout. The IXTOC I blowout, Bay of Campeche. Environmental Sci. Technol, 16(2), 67-74.

Bopp, L., C. Le Quere, M. Heimann, and A. C. Manning. 2002. Climate-induced oceanic oxygen fluxes: Implications for the contemporary budget. Global Biogeochemical Cycles, 16(0), 10.1029/2001GB001445.

Broecker, W. S. 1974. Chemical Oceanography. Hartcourt Brace Jovanovich Inc., New York, NY. 214 pp.

Broecker, W. S. and T.-H. Peng. 1982. Tracers in the Sea. ELDIGIO Press, Lamont-Doherty Geological Observatory, Columbia University, New York. 690 pp.

Brooks, D. A. 1983. The wake of Hurricane Allen in the western Gulf of Mexico. J. Phys. Oceanogr., 13(1), 117–129.

Buerkert, T. P. 1997. Barium in water and foraminiferal shells: indicators of present and past oceanographic conditions in the Gulf of Mexico. Ph.D. Thesis. Louisiana State University, Baton Rouge, LA.

Bunge, L., J. Ochoa, A. Badan, J. Candela, and J. Sheinbaum. 2002. Deep flows in the Yucatan Current and their relation to changes in the Loop Current extension. J. Geophys. Res., 107(C12), 3233, doi:10.1029/2001JC001256.

Carder, K. L., K. A. Fanning, P. R. Betzer, and V. Maynard. 1977. Dissolved silica and the circulation in the Yucatan Strait and deep eastern Gulf of Mexico. Deep-Sea Res. 24, 1149-1160.

Carney, R. S. 2001. Management applicability of contemporary deep-sea ecology and reevaluation of Gulf of Mexico studies. U.S. Dept. of the Interior, Minerals Management Service, Gulf of Mexico OCS Region, New Orleans. OCS Study MMS 2001-095. 174 pp.

Carpenter, J. H. 1965a. The accuracy of the Winkler method for dissolved oxygen analysis. Limnol. Oceanogr., 10, 135-140.

Carpenter, J. H. 1965b. The Chesapeake Bay Institute technique for the Winkler dissolved oxygen method. Limnol Oceanogr., 10, 141-143.

Carpenter, J. H. 1966. New measurements of oxygen solubility in pure and natural waters. Limnol Oceanogr., 11, 265-277.

Chapman, P. 1985. Oil concentrations in seawater following dispersion with and without the use of chemical dispersants. Special Report 2. Sea Fisheries Research Institute, Cape Town, South Africa. 23 pp.

Chester, R. 2003. Marine Geochemistry. Chapman and Hall, London, U.K.

CICESE and T. M. Mitchell. 2002. Yucatan inflow measurement program: Final Report, June 2002. DEEPSTAR Report 4804.

Cochrane, J.D. 1963. Yucatan Current. In Oceanography and Meteorology of the Gulf of Mexico, Annual Report, 1 May 1962-30 April 1963. Texas A&M University, Department of Oceanography, College Station, TX. Tech. Rpt. Ref. 63-18A. pp. 6-11.

Continental Shelf Associates, Inc. 2000. Deepwater Gulf of Mexico environmental and socioeconomic data search and literature synthesis. Volume I: Narrative Report, OCS Study MMS 2000-049. Volume II: Annotated bibliography, OCS Study MMS 2000-050. U.S. Dept. of the Interior, Minerals Management Service, Gulf of Mexico OCS Region, New Orleans.

Cooper, C., G. Z. Forristall, and T. M. Joyce. 1990. Velocity and hydrographic structure of two Gulf of Mexico warm-core rings. J. Geophys. Res., 95(C2), 1663–1679.

Craig, H. and T. Hayward. 1987. Oxygen supersaturation in the ocean: biological versus physical contributions. Science, 235 (4785), 199-202.

Crawford, T. G., G. L. Burgess, C. J. Kinler, M. T. Prendergast, and K. M. Ross. 2003. Estimated oil and gas reserves, Gulf of Mexico Outer Continental Shelf, December 31, 2000. U.S. Department of the Interior, Minerals Management Service, Gulf of Mexico OCS Region, New Orleans, LA. OCS Study MMS 2003-050, 30 pp.

Culberson, C.H. 1991. Dissolved oxygen. In WOCE Operations Manual, Volume 3, Part 3.1.3, WHP Operations and Methods - July 1991. WHP Office Report, WHPO 91-1, WOCE Report No. 68/91, November 1994, Revision 1, WOCE Hydrographic Programme Office, Woods Hole Oceanographic Institution, Woods Hole, MA. [available electronically in PDF at http://whpo.ucsd.edu/manuals.htm]

Culberson, C.H., G. Knapp, M.C. Stalcup, R.T. Williams, and F. Zemlyak. 1991. A comparison of methods for the determination of dissolved oxygen in seawater. WHPO Publication 91-2, WOCE Report 73/91, August 1991, WOCE Hydrographic Programme Office, Woods Hole Oceanographic Institution, Woods Hole, MA.

De Beukelaer, S.M., I.R. MacDonald, N.L. Guinasso Jr., and J.A. Murray. 2003. Distinct side-scan sonar, RADARSAT SAR, and acoustic profiler signatures of gas and oil seeps on the Gulf of Mexico slope. DOI 10.1007/s00367-003-0139-9. Geo-Mar. Lett., 23, 177-186.

DiMarco, S. F. , M. K. Howard, and A. E. Jochens. 2001. Deepwater Gulf of Mexico Historical Physical Oceanography Data Report. TAMU Oceanography Technical Report No. 01-01-D. Texas A&M University, Department of Oceanography, College Station, TX. 203 pp.

Drummond, K. H., compiler. 1956. Oceanographic data, Gulf of Mexico, cruises 55-1A, 55-1B, 55-1C and 55-3 of the A. A. Jakkula. A&M Project 24, Reference 56-33D, December 1956. The A&M College of Texas, Department of Oceanography and Meteorology, College Station, TX.

Drummond, K. H., compiler. 1957. Oceanographic data, Gulf of Mexico, cruises 55-4, 55-7, and 55-8 of the A. A. Jakkula. A&M Project 24, Reference 57-5D, February 1957. The A&M College of Texas, Department of Oceanography and Meteorology, College Station, TX.

Elliott, B.A. 1982. Anticyclonic rings in the Gulf of Mexico. J. Phys. Oceanogr. 12, 1292-1308.

El-Sayed, S. Z., W. M. Sackett, L. M. Jeffrey, A. D. Fredericks, R. P. Saunders, P. S. Conger, G. A. Fryxell, K. A. Steindinger, and S. A. Earle. 1972. Chemistry, primary productivity, and benthic algae of the Gulf of Mexico. In V. C. Bushnell (ed.) Folio 22, Serial Atlas of the Marine Environment, American Geographical Society.

Energy Information Administration. 2002. U.S. Crude Oil, Natural Gas, and Natural Gas Liquids Reserves 2002 Annual Report, December 2002. Energy Information Administration, Office of Oil and Gas, U.S. Department of Energy, Washington, DC. DOE/EIA–0216(2002), 170 pp.

Energy Information Administration. 2004. International energy annual 2002. Energy Information Administration, U.S. Department of Energy, Table 8.1, world crude oil and natural gas reserves, Jan 1, 2003, http://www.eia.doe.gov.pub/inernational/iea202/table81.xls, posted 8 Mar 2004.

Fiadero, M.E., and H. Craig. 1978. Three-dimensional modeling of tracers in the deep Pacific Ocean: I. Salinity and oxygen. J. Mar. Res. 36, 323 – 355.

Fratantoni, P.S, T. N. Lee, G. P. Podesta, F. Muller-Karger. 1998. The influence of Loop Current perturbations on the formation and evolution of Tortugas eddies in the southern Straits of Florida. J. Geophys. Res., 103(C11), 24759-24779.

Gage, J.D., L.A. Levin, and G.A. Wolff, eds. 2000. Benthic processes in the deep Arabian Sea: Biogeochemistry, biodiversity and ecology. Deep-Sea Res., Part II, 47(Nos. 1-2), 1-7.

Gibbs, C. F. 1976. Methods and interpretation in measurement of oil biodegradation rate. Biodegradation of materials, Vol. 3, Applied Science Publications, pp. 127-140.

Hamilton, P. 1990. Deep currents in the Gulf of Mexico. J. Phys. Oceanogr. 20, 1087-1104.

Hamilton, P. 1992. Lower continental slope cyclonic eddies in the central Gulf of Mexico, J. Geophys. Res., 97 (C2), 2185-2200.

Hamilton, P., and A. Lugo-Fernandez. 2001. Observations of high-speed deep currents in the northern Gulf of Mexico. Geophys. Res. Lett., 28, 2867-2870.

Hamilton, P., E. Waddell, and R. Wayland. 1997. Chapter 2 - The Physical Environment. In Northeastern Gulf of Mexico Coastal and Marine Ecosystem Program, Data Search and

Synthesis; Synthesis Report, SAIC. U.S. Department of the Interior, USGS, Biological Resources Division, USGS/BRD/CR-1997-0005 and Minerals Management Service, Gulf of Mexico OCS Region, New Orleans, LA, OCS Study MMS 96-0014. 313 pp.

Hamilton, P., G. S. Fargion, and D. C. Biggs. 1999. Loop Current eddy paths in the western Gulf of Mexico. J. Phys. Oceanogr., 29(6), 1180–1207.

Hamilton, P., T. J. Berger, and W. Johnson. 2002. On the structure and motions of cyclones int he northern Gulf of Mexico. J. Geophys. Res., 107(C12), 3208, doi:10.1029/1999JC000270.

Hansen, D. V, and R. L. Molinari. 1979. Deep currents in the Yucatan Strait. J. Geophys. Res., 84, 359-362.

Hedges, J. I., J. A. Baldcok, Y. Gelinas, C. Lee, M. L. Peterson, and S. G. Wakeham. 2002. The biochemical and elemental compositions of marine plankton: A NMR perspective. Marine Chemistry, 78, 47-63.

Hofman, E. E., and S. J. Worley. 1986. An investigation of the circulation of the Gulf of Mexico. J. Geophys. Res., 91(C12), 14221-14236.

Huh, O. K., W. J. Wiseman, Jr., and L. J. Rouse, Jr. 1981. Intrusion of Loop Current waters onto the west Florida continental shelf. J. Geophys. Res., 86(C5), 4186–4192.

Ichiye, T. 1972. Circulation changes caused by hurricanes. In L. R. A. Capurro and J. L. Reid, eds., Contributions on the Physical Oceanography of the Gulf of Mexico. Gulf Publishing Company, Houston, TX. pp. 229-257.

Ichiye, T., H-H. Kuo, and M. R. Carnes. 1973. Assessment of currents and hydrography of the eastern Gulf of Mexico. Contribution Number 601. Department of Oceanography, Texas A&M University, College Station, TX. 343 pp.

Jahnke, R.A., and G. A. Jackson. 1992. The spatial distribution of sea floor oxygen consumption in the Atlantic and Pacific oceans. In G. T. Rowe and V. Pariente, eds., Deep-Sea Food Chains and the Global Carbon Cycle. Kluwer Academic Publishers, Netherlands. pp 295-307.

Jenkins, W. J. 1982. Oxygen utilization rates in North Atlantic subtropical gyre and primary production in oligotrophic systems. Nature, 300, 246-248.

Jenkins, W. J. 1998. Studying subtropical thermocline ventilation and circulation using tritium and [3]He. J. Geophys. Res., 103(C8), 15817-15831.

Jochens, A. E., and W. D. Nowlin, Jr. 1998. Northeastern Gulf of Mexico Chemical Oceanography and Hydrography Study between the Mississippi Delta and Tampa Bay, Annual Report: Year 1. OCS Study MMS 98-0060. U. S. Department of the Interior, Minerals Management Service, Gulf of Mexico OCS Region, New Orleans, LA. 126 pp.

Jochens, A. E., and W. D. Nowlin, Jr. 1999. Northeastern Gulf of Mexico Chemical Oceanography and Hydrography Study, Annual Report: Year 2. OCS Study MMS 99-0054. U. S. Department of the Interior, Minerals Management Service, Gulf of Mexico OCS Region, New Orleans, LA. 123 pp.

Jochens, A. E., and W. D. Nowlin, Jr. 2000. Northeastern Gulf of Mexico Chemical Oceanography and Hydrography Study, Annual Report: Year 3. OCS Study MMS 2000-078. U. S. Department of the Interior, Minerals Management Service, Gulf of Mexico OCS Region, New Orleans, LA. 89 pp.

Jochens, A. E., D. A. Wiesenburg, L. E. Sahl, C. N. Lyons, and D. A. DeFreitas. 1998. LATEX Shelf Data Report: Hydrography, April 1992 through November 1994. TAMU Oceanography Tech. Rpt. No. 96-6-T. Texas A&M University, College Station TX. 6 volumes. [Available with the data through NODC on CD-ROM NODC-92, Texas-Louisiana Shelf Circulation and Transport Processes Study, Hydrography, Drifters, ADCP, and Miscellaneous Sensors Data and Reports, 1992-1994.]

Jochens, A. E., S. F. DiMarco, W. D. Nowlin, Jr., R. O. Reid, and M. C. Kennicutt II. 2002. Northeastern Gulf of Mexico Chemical Oceanography and Hydrography Study: Synthesis Report. U.S. Dept. of the Interior, Minerals Management Service, Gulf of Mexico OCS Region, New Orleans, LA. OCS Study MMS 2002-055. 586 pp.

Johansen, O., H. Rye, and C. Cooper. 2002. DeepSpill–Field study of a simulated oil and gas blowout in deep water. (in press).

Karl, D. M., P. LaRock, J. W. Morse and W. Sturges. 1976. Adenosine triphosphate in North Atlantic Ocean sediments and its relationship to the oxygen minimum. Deep-Sea Res., 23, 81-88.

Kennicutt, M. C, II. 2000. Chapter 5 – Chemical Oceanography, 123-140. In Continental Shelf Associates, Inc. Deepwater Gulf of Mexico Environmental and Socioeconomic Data Search and Literature Synthesis. Volume I: Narrative Report. U.S. Department of the Interior, Minerals Management Service, Gulf of Mexico OCS Region, New Orleans, LA. OCS Study MMS 2000-049. 340 pp.

Key, R. M. 1981. Examination of abyssal sea floor and near-bottom mixing processes using ^{226}Ra and ^{222}Rn. Ph.D. dissertation, Texas A&M University, College Station, TX. 227pp.

Kinghorn, R. R. F. 1983. An Introduction to the Physics and Chemistry of Petroleum. John Wiley & Sons, New York. 420 pp.

Le Bureau Du Conseil Service Hydrographique. 1936. Bulletin Hydrographique Pour L'Année 1935. Conseil Permanent International Pour L'Exploration de la Mer, Charlottenlund Slot, Danemark.

Leipper, D. F., project supervisor. 1956. Oceanographic survey of the Gulf of Mexico: Physical and meteorological data, cruises 54-2, 54-9, and 54-10 of the A. A. Jakkula. A&M Project 24, Reference 56-7D, March 1956. The A&M College of Texas, Department of Oceanography and Meteorology, College Station, TX.

Leipper, D. F. 1968a. Hydrographic station data, Gulf of Mexico: August-November Nansen Casts, 1965-1967. Reference 68-13T, August 1968. Department of Oceanography, Texas A&M University, College Station, TX.

Leipper, D. F. 1968b. Hydrographic station data, Gulf of Mexico: February-March Nansen Casts, 1965-1968. Reference 68-15T, August 1968. Department of Oceanography, Texas A&M University, College Station, TX.

Leipper, D. F. 1968c. Hydrographic station data, Gulf of Mexico: August 17 - September 5, 1968 Nansen Casts and STD. Reference 68-17T, August 1968. Department of Oceanography, Texas A&M University, College Station, TX.

Leipper, D. F. 1970. A sequence of current patterns in the Gulf of Mexico. J. Geophys. Res., 75(3), 637-657.

Libes, S. M. 1992. An Introduction to Marine Biogeochemisty. John Wiley & Sons, Inc., New York. 734 pp.

Maul, G. A. 1976. The 1972-1973 cycle of the Gulf Loop Current, Part II: Mass and salt balances of the basin. In CICAR-II, symposium on Progress in marine Research in the Caribbean and Adjacent Regions, Caracas, Venezuela. FAO Fisheies Report No. 200 Supplement, pp. 597-619.

Maul, G. A., D. A. Mayer, and S. R. Baig. 1985. Comparisons between a continuous 3-year current-meter observations at the sill of the Yucatan Strait, satellite measurements of Gulf Loop Current A area, and regional sea level. J. Geophys. Res., 90(C5), 9089–9096.

MacDonald, I. R. 1992. Chemosynthetic Ecosystems Study Literature Review and Data Synthesis, Volume I: Executive Summary. Minerals Management Service, Gulf of Mexico OCS Regional Office, U. S. Dept. of the Interior, New Orleans, LA. MMS OCS Study 1992-033. 32 pp.

MacDonald, I. R., ed. 2002. Stability and Change in Gulf of Mexico Chemosynthetic Communities. Volume II: Technical Report. Minerals Management Service, Gulf of Mexico OCS Regional Office, U.S. Dept. of the Interior, New Orleans, LA. OCS Study MMS 2002-036. 456 pp.

MacDonald, I .R., and W.W. Schroeder. 1993. Chemosynthetic Ecosystems Studies Interim Report. Minerals Management Service, Gulf of Mexico OCS Regional Office, U.S. Dept. of the Interior, New Orleans, LA. Study MMS 93-0032. 110 pp.

MacDonald, I .R., W.W. Schroeder, and J.M. Brooks. 1995. Chemosynthetic Ecosystems Study Final Report, Volume I: Technical Report. Minerals Management Service, Gulf of Mexico OCS Regional Office, U.S. Dept. of the Interior, New Orleans, LA. Study MMS 95-0022. 338 pp.

Mathews, T. D., A. D. Fredericks, and W. M. Sackett. 1973. The geochemistry of radiocarbon in the Gulf of Mexico. In Radioactive Contamination of the Marine Environment. IAEA Symp. IAEA-SM-158/48. 725-734

McLellan, H. J. 1959. Data report for two IGY cruises. A&M Projects 137 and 24, Reference 59-15D, April 1959. The A&M College of Texas, Department of Oceanography and Meteorology, College Station, TX.

McLellan, H. J. 1960. The waters of the Gulf of Mexico as observed in 1958 and 1959. A&M Projects 205 and 24, Reference 60-14T, September 1960. The A&M College of Texas, Department of Oceanography and Meteorology, College Station, TX. 61 pp.

McLellan, H. J., and W. D. Nowlin, Jr. 1962. The waters of the Gulf of Mexico as observed in February and March 1962: Data Report. A&M Project 286, Reference 62-16D, September 1962. The A&M College of Texas, Department of Oceanography and Meteorology, College Station, TX.

McLellan, H. J., and W. D. Nowlin, Jr. 1963. Some features of the deep water in the Gulf of Mexico. J. Mar. Res., 21 (3), 233–245.

Metcalf, W. G. 1976. Caribbean-Atlantic water exchange through the Anegarla-Jungfern Passage. J. Geophys. Res., 81, 6401-6409.

Millero, F. J. 1996. Chemical Oceanography (2nd Edition). CRC Press Inc., Boca Raton, Florida. 469 pp.

Mitchell, R., I. R. MacDonald, and K. A. Kvenvolden. 1999. Estimation of total hydrocarbon seepage into the Gulf of Mexico based on satellite remote sensing images. Transactions, American Geophysical Union 80(49), Ocean Sciences Meeting Supplement, OS242.

Molinari, R. L., and J. Morrison. 1988. The separation of the Yucatan Current from the Campeche Bank and the intrusion of the Loop Current into the Gulf of Mexico. J. Geophys. Res., 93 (C9), 10645-10654.

Molinari, R. L., J. F. Festa, and D. W. Behringer. 1978. The circulation in the Gulf of Mexico derived from estimated dynamic height fields. J. Phys. Oceanogr., 8, 987-996.

Morrison, J. M., and W. D. Nowlin, Jr. 1977. Repeated nutrient, oxygen, and density sections through the Loop Current. J. Mar. Res., 35(1), 105–128.

Morrison, J. M., and W. D. Nowlin, Jr. 1982. General distribution of water masses within the eastern Caribbean Sea during the winter of 1972 and fall of 1973. J. Geophys. Res., 87(C6), 4207-4229.

Morrison, J. M., W. J. Merrell, Jr., and W. D. Nowlin, Jr. 1973. The waters of the eastern Gulf of Mexico as observed during May 1972. I. Data collected aboard the R/V Alaminos: Data Report. Reference 73-10-T, May 1973. Department of Oceanography, Texas A&M University, College Station, TX.

Morrison, J. M., W. J. Merrell, Jr., R. M. Key, and T. C. Key. 1983. Property distributions and deep chemical measurements within the western Gulf of Mexico. J. Geophys. Res., 88(C4), 2601-2608.

National Research Council (NRC). 1975. Petroleum in the Marine Environment: Workshop on Inputs, Fates, and the Effects of Petroleum in the Marine Environment, May 21-25, 1973. National Academy Press, Washington D.C. 107 pp.

National Research Council (NRC). 1985. Oil in the Sea: Inputs, Fates, and Effects. National Academy Press, Washington D.C. 601 pp.

National Research Council (NRC). 2003. Oil in the Sea III: Inputs, Fates, and Effects. National Academy Press, Washington D.C. 280 pp.

Neff, J. M., S. McKelvie and R. C. Ayers, Jr. 2000. Environmental impacts of synthetic based drilling fluids. Report prepared for MMS by Robert Ayers & Associates, Inc. August 2000. U.S. Department of the Interior, Minerals Management Service, Gulf of Mexico OCS Region, New Orleans, LA. OCS Study MMS 2000- 064. 118 pp.

Nowlin, W. D., Jr. 1972. Winter Circulation Patterns and Property Distributions, in Contributions on the Physical Oceanography of the Gulf of Mexico. L. R. A. Capurro and J. L. Reid, eds. Texas A&M University Oceanographic Studies, Volume 2, pp. 3–51. Gulf Publishing Co., Houston, 288 pp.

Nowlin, W. D., Jr., and H. McLellan. 1967. A characterization of the Gulf of Mexico waters in winter. J. Mar. Res., 25 (1), 29-59.

Nowlin, W. D., Jr., and C. A. Parker. 1974. Effects of a cold-air outbreak on shelf waters of the Gulf of Mexico. J. Phys. Oceanogr., 4 (3). 467-486.

Nowlin, W. D., Jr., D. F. Paskausky, and H. J. McLellan. 1969. Recent dissolved-oxygen measurements in the Gulf of Mexico deep waters. J. Mar. Res., 27 (1), 39-44.

Nowlin, W. D., Jr., A. E. Jochens, R. O. Reid, and S. F. DiMarco. 1998a. Texas-Louisiana Shelf Circulation and Transport Processes Study: Synthesis Report, Volume I: Technical Report, OCS Study MMS 98-0035. U.S. Department of the Interior, Minerals Management Service, Gulf of Mexico OCS Region, New Orleans, LA, 502 pp.

Nowlin, W. D., Jr., A. E. Jochens, R. O. Reid, and S. F. DiMarco. 1998b. Texas-Louisiana Shelf Circulation and Transport Processes Study: Synthesis Report, Volume II: Appendices, OCS Study MMS 98-0036. U.S. Department of the Interior, Minerals Management Service, Gulf of Mexico OCS Region, New Orleans, LA. 288 pp.

Nowlin, W. D., Jr., A. E. Jochens, S. F. DiMarco, and R. O. Reid. 2000. Chapter 4: Physical Oceanography, in Continental Shelf Associates, Inc., Deepwater Gulf of Mexico Environmental and Socioeconomic Data Search and Literature Synthesis. Volume I: Narrative Report. pp. 61-121. U.S. Department of the Interior, Minerals Management Service, Gulf of Mexico OCS Region, New Orleans, LA, OCS Study MMS 2000-049. 340 pp.

Nowlin, W. D., Jr., A. E. Jochens, S. F. DiMarco, R. O. Reid, and M. K. Howard. 2001. Deepwater Physical Oceanography Reanalysis and Synthesis of Historical Data: Synthesis Report. OCS Study MMS 2001-064, U.S. Department of the Interior, Minerals Management Service, Gulf of Mexico OCS Region, New Orleans, LA. 528 pp.

Ochoa, J., J. Sheinbaum, A. Badan, J. Candela, and D. Wilson. 2001. Geostrophy via potential vorticity inversion in the Yucatan Channel. J. Mar. Res., 59(5), 725-747.

Owens, W. B., and R. C. Millard Jr. 1985. A new algorithm for CTD oxygen calibration. J. Physical Oceanogr., 15, 621-631

Paluszkiewicz, T., L. A. Atkinson, E. S. Posmentier, and C. R. McClain. 1983. Observations of a Loop Current frontal eddy intrusion onto the west Florida shelf. J. Geophys. Res., 88(C14), 9639–9651.

Pilson, M. E. Q. 1998. An Introduction to the Chemistry of the Sea. Prentice-Hall, Inc., Upper Saddle River, New Jersey. 431 pp.

Rabalais, N. N., and R. E. Turner. 2001. Hypoxia in the northern Gulf of Mexico: description, causes and change. In N. N. Rabalais and R. E. Turner, eds, Coastal hypoxia: consequences for living resources and ecosystems. Coastal and Estuarine Studies 58. American Geophysical Union, Washington, D.C. pp. 1-36.

Redfield, A. C, B. H. Ketchum, and F. A. Richards. 1963. The influence of organisms on the composition of sea water. In M. N. Hill, ed., The Sea: Ideas and observations on progress in the study of the seas. Volume 2: The composition of sea-water, comparative and descriptive oceanography. John Wiley & Sons, New York. pp. 26-77.

Richards, F. A. 1965. Chapter 6: Dissolved Gases Other than Carbon Dioxide. In J. P. Riley and G. Skirrow. Chemical Oceanography, Volume 1, 1st ed., Academic Press, Ipswich, Suffolk, UK. pp 197-225.

Riley, G. A. 1951. Oxygen, phosphate, and nitrate in the Atlantic Ocean. Bull. Bingham oceanographic Collection. XII (article 1), 1- 126.

Riley, J. P., and R. Chester. 1971. Introduction to Marine Chemistry. Academic Press, Inc. Orlando, FL. 465 pp.

Rogers, A. D. 2000. The role of the oceanic oxygen minima in generating biodiversity in the deep sea. Deep-Sea Res., 47 (no. 1-2), 119-148.

Rowe, G. T. 2005. Deepwater program: Northern Gulf of Mexico Continental Slope Habitat and Benthic Ecology, Draft Final Report. U.S. Department of the Interior, Minerals Management Service, Gulf of Mexico OCS Region, New Orleans, Louisiana. OCS Study MMS 2005-xxx, yy pp.

Rowe, G. T, A. Lohse, F. Hubbard, G. S. Boland, E. E. Briones, and J. Deming. 2003. Preliminary trophodynamic carbon budget for the Sigsbee deep benthos, northern Gulf of Mexico. Am. Fisheries Soc. Symp., 36, 225-238.

Sahl, L. E., D. A. Wiesenburg, and W. J. Merrell. 1997. Interactions of mesoscale features with Texas shelf and slope waters. Cont. Shelf Res., 17(2), 117-136.

Schmitz, W. J., and W. S. Richardson. 1968. On the transport of the Florida Current. Deep-Sea Res., 15, 679-693.

Science Applications International Corporation (SAIC). 1986. Gulf of Mexico physical oceanography program, Final Report: Years 1 and 2, Volume II: Technical Report. OCS Study MMS 85-0094, U.S. Dept. of the Interior, Minerals Management Service, Gulf of Mexico OCS Region, New Orleans, 403 pp.

Science Applications International Corporation (SAIC). 1987. Gulf of Mexico physical oceanography program, Final Report: Year 4, Volume II: Technical Report. OCS Study MMS 87-0007, U.S. Dept. of the Interior, Minerals Management Service, Gulf of Mexico OCS Region, New Orleans, 315 pp.

Science Applications International Corporation (SAIC). 1988. Gulf of Mexico physical oceanography program, Final Report: Year 3, Volume II: Technical Report. OCS Study MMS 88-0046, U.S. Dept. of the Interior, Minerals Management Service, Gulf of Mexico OCS Region, New Orleans, 266 pp.

Science Applications International Corporation (SAIC). 1989. Gulf of Mexico physical oceanography program, Final Report: Year 5, Volume II: Technical Report. U.S. Dept. of the Interior, Minerals Management Service, Gulf of Mexico OCS Region, New Orleans, LA. OCS Study MMS 89-0068. 333 pp.

Sheinbaum, J., J. Candela, A. Badan, and J. Ochoa. 2002. Flow structure and transport in the Yucatan Channel. Geophys. Res. Letters, 29(3), 10.1029/2001GL013990.

Shiller, A. M. 1999. An overview of the marine chemistry of the Gulf of Mexico. In H. Kumpf, K. Steidinger, and K. Sherman, eds., The Gulf of Mexico Large Marine Ecosystem, Blackwell Science, Inc., Malden, MA. pp. 132-148.

Strickland, J. D. H., and T. R. Parsons. 1972. A Practical Handbook of Seawater Analysis. Bulletin 167 (Second edition). Fisheries Research Board of Canada, Ottawa.

Sturges, W. 1993. The annual cycle of the western boundary current in the Gulf of Mexico. J. Geophys. Res., 98(C10), 18053-18068.

Sturges, W. 1994. The frequency of ring separations from the Loop Current. J. Phys. Oceanogr., 24, 1647-1651.

Sturges, W., and R. Leben. 2000. Frequency of ring separations from the Loop Current in the Gulf of Mexico: a revised estimate. J. Phys. Oceanogr., 30, 1814-1819.

Sturges, W., J. Evans, S. Welsh, and W. Holland. 1993. Separation of warm-core rings in the Gulf of Mexico. J. Phys. Oceanogr., 23, 250-268.

Sturges, W., E. Chassignet, and T. Ezer. 2004. Strong mid-depth currents and a deep cyclonic gyre in the Gulf of Mexico. U.S. Dept. of the Interior, Minerals Management Service, Gulf of Mexico OCS Region, New Orleans, LA. OCS Study MMS 2004-040. 89 pp.

Takahashi, T., W. S. Broecker, and S. Langer. 1985. Redfield ratio based on chemical data from isopycnal surfaces. J. Geophys. Res., 90(C4), 6908-6924.

Tréguer, P., D. M. Nelson, A. J. Van Bennekom, D. J. DeMaster, A. Leynaert, and B. Queguiner. 1995. The silica balance in the world ocean: a reestimate. Science, 268(5209), 375-379.

Vázquez de la Cerda, A. M. 1993. Bay of Campeche cyclone. Ph.D. Dissertation. College Station, TX, Texas A&M University. 91 pp.

Vázquez de la Cerda, A. M., R. O. Reid, S. F. DiMarco, and A. E. Jochens. 2005. Bay of Campeche circulation: an update. In W. Sturges and A. Lugo-Fernandez, eds., New Developments in the Circulation of the Gulf of Mexico, Geophysical Monograph Series XX, American Geophysical Union, Washington, D.C. (in preparation).

Vidal, V. M., F. V. Vidal, A. F. Hernandez, E. Meza, and J. M. Perez-Molero. 1994. Baroclinic flows, transports, and kinematic properties in a cyclonic-anticyclonic-cyclonic ring triad in the Gulf of Mexico. J. Geophys. Res., 99(C4), 7571–7597.

Von Oudenhoven, J. A. C. M., V. Draper, G. P. Ebbon, P. D. Holmes, J. L. Nooyen. 1983. Characteristics of petroleum and its behavior at sea. CONCAWE's Oil Spill Clean-up Technology Task Force No. 8, CONCAWE, Den Haag. 47 pp.

Vukovich, F. M. 1995. An updated evaluation of the Loop Current's eddy shedding frequency. J. Geophys. Res., 100, 8655–8660.

Weatherly, G. 2004. Intermediate depth circulation in the Gulf of Mexico: PALACE Float Results for the Gulf of Mexico between April 1998 and March 2002. U.S. Department of the Interior, Minerals Management Service, Gulf of Mexico OCS Region, New Orleans, LA. OCS Study MMS 2004-13. 51 pp.

Welsh, S. E., and M. Inoue. 2000. Loop Current rings and the deep circulation in the Gulf of Mexico. J. Geophys. Res., 105, 16,951–16,959.

Welsh, S. E., and M. Inoue. 2002. Lagrangian study of circulation, transport, and vertical exchange in the Gulf of Mexico. U.S. Department of the Interior, Minerals Management Service, Gulf of Mexico OCS Region, New Orleans, LA. OCS Study MMS 2002-064. 51 pp.

Wennekens, M. P. 1959. Water mass properties of the Straits of Florida and related waters. Bulletin of Marine Science of the Gulf and Caribbean, 9 (1), 1-52.

Winkler, L. W. 1888. Ber. dtsch.chem. Ges., 21, 2843.

WOCE. 1991. WOCE Operations Manual, Volume 3, Part 3.1.3, WHP Operations and Methods - July 1991. WHP Office Report, WHPO 91-1, WOCE Report No. 68/91, November 1994, Revision 1, WOCE Hydrographic Programme Office, Woods Hole Oceanographic Institution, Woods Hole, MA. [available electronically in PDF at http://whpo.ucsd.edu/manuals.htm]

Wüst, G. 1964. Stratification and circulation in the Antillean-Caribbean basins. Part 1: Spreading and mixing of the water types. Columbia University Press. New York. 201 pp.

ZoBell, C. E. 1969. Microbial modification of crude oil in the sea. Proceeding of the Joint Conference on Prevention and Control of Oil Spills, New York, New York; pp. 317-326.

The Department of the Interior Mission

As the Nation's principal conservation agency, the Department of the Interior has responsibility for most of our nationally owned public lands and natural resources. This includes fostering sound use of our land and water resources; protecting our fish, wildlife, and biological diversity; preserving the environmental and cultural values of our national parks and historical places; and providing for the enjoyment of life through outdoor recreation. The Department assesses our energy and mineral resources and works to ensure that their development is in the best interests of all our people by encouraging stewardship and citizen participation in their care. The Department also has a major responsibility for American Indian reservation communities and for people who live in island territories under U.S. administration.

The Minerals Management Service Mission

As a bureau of the Department of the Interior, the Minerals Management Service's (MMS) primary responsibilities are to manage the mineral resources located on the Nation's Outer Continental Shelf (OCS), collect revenue from the Federal OCS and onshore Federal and Indian lands, and distribute those revenues.

Moreover, in working to meet its responsibilities, the **Offshore Minerals Management Program** administers the OCS competitive leasing program and oversees the safe and environmentally sound exploration and production of our Nation's offshore natural gas, oil and other mineral resources. The MMS **Minerals Revenue Management** meets its responsibilities by ensuring the efficient, timely and accurate collection and disbursement of revenue from mineral leasing and production due to Indian tribes and allottees, States and the U.S. Treasury.

The MMS strives to fulfill its responsibilities through the general guiding principles of: (1) being responsive to the public's concerns and interests by maintaining a dialogue with all potentially affected parties and (2) carrying out its programs with an emphasis on working to enhance the quality of life for all Americans by lending MMS assistance and expertise to economic development and environmental protection.